Planetary Accounting

Kate Meyer • Peter Newman

Planetary Accounting

Quantifying How to Live Within Planetary
Limits at Different Scales of Human Activity

 Springer

Kate Meyer
The Planetary Accounting Network
Auckland, New Zealand

Peter Newman
Curtin University
Western Australia, WA, Australia

ISBN 978-981-15-1445-6   ISBN 978-981-15-1443-2   (eBook)
https://doi.org/10.1007/978-981-15-1443-2

This Springer imprint is published by the registered company Springer Nature Singapore Pte Ltd.
The registered company address is: 152 Beach Road, #21-01/04 Gateway East, Singapore 189721, Singapore

Dedicated to Kate's children Patrick and Hazel and Peter's grandchildren Josie, Lilah and Benji. We hope planetary accounting will help to create a healthier planet for your generation.

Preface

There is a lot of science in this book, but a lot of other stuff too. We both know that science is really important to the issues we are addressing but each of us has lived in other realms of life, where we have learned a lot of "other stuff". I have a PhD in Chemistry, studied Environmental Science, and in 2018/2019 was named as Western Australian Scientist of the Year. Kate is a Mechanical Engineer and has just completed her PhD. So, we qualify to address the scientific issues, and we spent a lot of time working to make sure we had the science of Planetary Boundaries right. But why we feel this book has something that nobody else has done is that we have integrated the science of how the world is changing with how you change the world to make it better.

The scientists who produced the Planetary Boundary publications did a remarkable job. They had to bring so much together from so many varied scientific journals and make it into a sensible set of conclusions—very scary conclusions about the state of the planet. However, there has been a reasonable time that has passed since these publications and they were communicated very strongly in science media outlets, but it has not made much impression on the general public, businesses, or politicians. While some governments and businesses see the value in applying the Planetary Boundaries, this has taken considerable effort for each organization and the outcomes are typically incomplete and ad hoc. We think we know why and have written a book to enable us to see if our approach helps make it more possible to have a coherent approach about making our world better on all the Planetary Boundaries.

The reason we believe it has not been well understood beyond the first reaction of "this is terrible" is that the Planetary Boundaries are not in metrics that are comparable with human activity and decision-making. They are not even the same type of metrics; they are apples and oranges which cannot be compared. They also cannot be scaled into meaningful parameters where we can take responsibility at different levels. So, we have tried to fix this.

The reason we were able to see the issues and had an approach to fix it is because we have both worked with industry and with governments on policy-oriented environmental tools. These tools come from State of the Environment Reporting which has been set up at all levels of government and industry, and we also have been closely involved in local community groups and local governments where people are trying to work out what environmental responsibility means for them.

So, we have brought a different perspective to the Planetary Boundaries that is based on a number of other areas of expertise in the social sciences and humanities.

The other motivation behind this book is that we are parents and even grandparents in my case. You cannot look into the eyes of your own progeny without worrying about whether we have some hope for their future. This book is aimed at providing that hope.

If you do not think we are serious, then you should know that in order to get the science of Planetary Boundaries right, and to see if the scientists who did this amazing compilation of science can agree with the approaches we have taken to reframe them, Kate had to go and meet them all, mostly in Europe. She packed up her husband and two young children, bought a van and drove around Europe for 8 months, presenting seminars and having in-depth discussions with the scientists. She even had the van towed away at night once, leaving her and her family stranded with nowhere to sleep! This is real commitment. Check out Kate's blog if you want to know more https://airphd.wordpress.com/.

We are keen therefore that you read this and see what you can do about the suggested solutions. We need lots of suggestions for what can be done with the new tools we have created. And we are happy to recognize you in any further publications we do on the subject. So write to us, we are not hard to find at Curtin University or the Planetary Accounting Network https://www.planetaryaccounting.org/.

Western Australia, WA, Australia   Peter Newman

Contents

1   Introduction.   1
    1.1   The Journey   1
    1.2   A Guide to This Book   5
    1.3   Publications   7
    References.   7

Part I   Earth-System Science, Management Theory, and Environmental
         Accounting: The Disconnect

2   The Science of Anthropogenic Climate Change.   11
    2.1   Introduction   12
    2.2   Earth as a System.   13
    2.3   We Are at Risk of Changing the State of the Planet   14
          2.3.1   Understanding CO_2 and Temperature   15
    2.4   Is the Earth Getting Warmer?.   18
          2.4.1   1970s Predictions of an Imminent Ice Age   19
          2.4.2   1934: The Warmest Year on Record.   23
          2.4.3   The 1998–2015 Warming Hiatus   23
          2.4.4   Global Warming or the Urban Heat Island Effect   25
          2.4.5   Antarctic Sea Ice Is Increasing   25
    2.5   Human Activity Is the Main Driver for the Changes
          to the State of the Earth   26
          2.5.1   Is Climate Change Caused by the Sun?.   27
          2.5.2   Human Emissions of CO_2 Are Insignificant   27
    2.6   Climate Change Matters   29
    2.7   Conclusion.   31
    References.   31

3   The Holocene, the Anthropocene, and the Planetary Boundaries . . .   35
    3.1   Introduction   36
    3.2   Planetary Limits: A Brief History   36
          3.2.1   The Population-Technology-Lifestyle Nexus   41

3.3   The Holocene Epoch   42
3.4   The Anthropocene Epoch.   43
3.5   Planetary Limits for a Holocene-Like State:
      The Planetary Boundaries   45
3.6   Conclusion   48
References.   48

**4   Managing the Earth System: Why We Need
      a Poly-Scalar Approach**   53
4.1   Introduction   54
4.2   Theories of Commons Management   54
      4.2.1   Conventional Theories on Managing the Commons   55
      4.2.2   Modern Theories on Managing the Commons   57
4.3   Theories of Behaviour Change   61
4.4   Change Theory and Sustainability.   63
4.5   Poly-Scalar Management of the Earth System   64
      4.5.1   The Montreal Protocol: A Successful Example
              of Top-Down Global Governance or of
              a Poly-Scalar Approach?   66
4.6   Conclusion   68
References.   68

**5   Environmental Accounting, Absolute Limits,
      and Systemic Change**   73
5.1   Introduction   74
5.2   Systemic vs. Incremental Change   75
5.3   Environmental Impact Assessments: A History.   78
      5.3.1   Life-Cycle Assessments.   78
      5.3.2   Environmental Footprints   79
      5.3.3   The Benefits of Environmental Accounting.   81
      5.3.4   Production Versus Consumption Accounting   82
5.4   Absolute Sustainability   82
5.5   Carbon Accounting: Environmental Accounting
      with Absolute Limits   85
5.6   Conclusions   86
References.   86

**6   Resolving the Disconnect Between Earth-System Science,
      Management Theory, and Environmental Accounting**   89
6.1   Introduction   90
6.2   Adapting the Planetary Boundaries   91
      6.2.1   National Limits for Switzerland.   91
      6.2.2   National Limits for Sweden.   93
      6.2.3   National Limits for South Africa   93
      6.2.4   National Limits for the Netherlands   93
      6.2.5   Regional Limits for the EU   94
      6.2.6   The FB-ESA Framework   94

6.2.7 Connecting the PBs and LCA 95
6.2.8 Comparing the Adaptations of the Planetary Boundaries .. 95
6.3 Why the Planetary Boundaries Are Difficult to Scale 102
6.4 Conclusions: A Need for New Global Limits 104
References... 104

Part II   Developing Planetary Quotas

7   Translating the Planetary Boundaries into Planetary Quotas 109
7.1 Introduction 110
7.2 Methodologies for Multidisciplinary Research 112
   7.2.1 Post-normal Science. 112
   7.2.2 Integrative Research. 113
   7.2.3 Methodology for the Development of the Planetary
         Quotas and Planetary Accounting Framework. 113
7.3 Methods 114
   7.3.1 Interconnectivity 115
7.4 From PBs to PQs 117
7.5 Conclusions 119
References... 120

8   A Planetary Quota for Carbon Dioxide 121
8.1 Introduction 122
8.2 Background 122
   8.2.1 The Carbon Cycle 122
   8.2.2 Why Separate Carbon Dioxide from Other
         Greenhouse Gases? 124
8.3 An Indicator for Carbon Dioxide Emissions 127
8.4 The Limit 127
   8.4.1 Total Radiative Forcing 128
   8.4.2 Ocean Acidification 128
   8.4.3 Biosphere Integrity. 128
   8.4.4 CO_2 Concentration............................... 129
8.5 Discussion 131
8.6 Conclusion.. 133
References... 133

**9   A Quota for Agricultural GHG Emissions
    (Methane and Nitrous Oxide)** 137
9.1 Introduction 138
9.2 The Methane Cycle 139
9.3 Nitrous Oxide.. 140
   9.3.1 Laughing Gas..................................... 140
   9.3.2 Impacts.. 141
9.4 An Indicator for Me-NO 141
9.5 The Limit 141

9.6   Discussion   143
9.7   Conclusions   144
References.   144

10 **A Quota for Forestland**.   147
10.1   Introduction   147
10.2   Background   148
10.3   The Importance of Forests   151
10.4   An Indicator for Forestland   152
10.5   The Limit   153
10.6   Discussion   154
10.7   Conclusions   154
References.   155

11 **A Quota for Ozone-Depleting Substances**   157
11.1   Introduction   157
11.2   Background   159
11.2.1   Montreal Protocol.   161
11.3   The Limit   163
11.4   Discussion   164
11.5   Conclusions   164
References.   165

12 **A Quota for Aerosols**.   167
12.1   Introduction   168
12.2   Background   169
12.3   Measuring Aerosols   170
12.4   Equivalent Aerosol Optical Depth   172
12.4.1   Calculating AODe   172
12.5   The Limit   173
12.5.1   Radiative Forcing.   174
12.5.2   Air Pollution.   175
12.5.3   The PQ for Aerosols.   176
12.6   Discussion   176
12.7   Conclusions   177
References.   178

13 **A Quota for Water**   181
13.1   Introduction   182
13.2   Background   182
13.3   A Global Problem   184
13.3.1   Weighting Water to Manage Regionality   186
13.4   The Indicator   189
13.4.1   Green Versus Blue   189
13.4.2   Gross Water Versus Net Water   191
13.4.3   Grey Water and Novel Entities.   191
13.5   The Limit   193

13.6   Discussion   193
13.7   Conclusion   195
References   195

14   A Quota for Nitrogen   197
14.1   Introduction   197
14.2   Background   198
       14.2.1   The Natural Nitrogen Cycle   198
       14.2.2   Human Use of Nitrogen   200
       14.2.3   The Impacts of Nitrogen   201
14.3   The Indicator   202
14.4   The Limit   203
14.5   Discussion   204
14.6   Conclusions   204
References   205

15   The Phosphorus Quota   207
15.1   Introduction   208
15.2   The Phosphorus Cycle   208
15.3   Human Use of Phosphorus: A Brief History   209
       15.3.1   The (Human) Phosphorous Cycle   212
       15.3.2   Phosphorus: A Nonrenewable Resource   214
15.4   An Indicator for Phosphorus   216
15.5   The Limit   217
15.6   Discussion   217
15.7   Conclusions   217
References   218

16   The Biodiversity Quota   221
16.1   Introduction   221
16.2   Background   222
       16.2.1   Biodiversity Management   225
       16.2.2   Measuring Biodiversity Health   226
16.3   The Indicator   228
       16.3.1   A Land-Based Proxy Indicator   233
16.4   The Limit   235
16.5   Discussion   235
16.6   Conclusion   236
References   236

17   The Imperishable Waste Quota   241
17.1   Introduction   241
17.2   Background   242
17.3   A Planetary Boundary with No Boundary   243
17.4   An Indicator for Novel Entities   244
17.5   A Limit   245

17.6   Discussion 245
17.7   Conclusion 246
References.. 246

Part III   A New Paradigm of Environmental Management

18   The Planetary Accounting Framework........................ 251
18.1   Introduction 252
18.2   The Planetary Quotas................................... 253
18.3   The Planetary Accounting Framework 256
       18.3.1   Bottom-Up..................................... 256
       18.3.2   Top-Down...................................... 260
       18.3.3   The Impact Balance Sheet 264
       18.3.4   Potential Applications 264
18.4   Discussion 267
       18.4.1   Timeframe 267
       18.4.2   Comparing Quotas 267
       18.4.3   The Quotas Are a Moving Target Not a Static Value 268
       18.4.4   Global vs Regional Limits and Impacts: An Issue
                of Scale 268
18.5   Strengths and Weaknesses 269
18.6   Ongoing and Future Work 270
       18.6.1   The Planetary Accounting Network................ 272
18.7   Concluding Remarks 274
References.. 277

Chapter 1
Introduction

Abstract Nine Planetary Boundaries were determined by a group of Earth-system scientists in 2009 but little social and political change towards address them has happened. Our book shows why and how this can be fixed through creating Planetary Accounting based on three social science theories. This introductory chapter provides an overview of how this will be done.

1.1 The Journey

Our planet is changing. One-in-one-hundred-year storms are becoming the norm. News articles about imminent species extinctions no longer feel shocking. They are expected. Images of oceans full of plastic litter social media. An island near Hawaii has been taken off the map as it has been engulfed by rising sea levels. The chemical composition of our atmosphere has not previously been experienced by humankind.

In 2009, 28 internationally renowned Earth-system scientists led by Johan Rockström at the Stockholm Resilience Centre (Rockström et al. 2009) identified nine Planetary Boundaries (PBs)—critical global environmental limits within which the risk of fundamentally changing the state of the planet is low: a "safe operating space" for humankind. If human activity pushes the planet beyond these limits, we are at risk of changing the state of the planet. The alternative state, beyond these limits, is uncertain, but it is likely to be hotter, less stable, and less favourable to humankind. Four of the PBs have been transgressed, the PBs for climate change, biogeochemical flows, land use, and biodiversity loss (Steffen et al. 2015). The situation is urgent. The risk that human activity will irrevocably change the state of the planet is high. We ought to manage human activity such that we can live well within the safe operating space. The problem is how?

In the face of this global environmental crisis, it can be hard to imagine what you could do to help. The PBs convey important information about the health of the planet. However, they do not tell us what to do. They are limits for the environment, not for people. They cannot be easily related to human activities. Nor do they make sense at smaller scales. Scientists and policy-makers want to translate the PBs into policy, to use them as the basis for managing human impacts on the

© Springer Nature Singapore Pte Ltd. 2020
K. Meyer, P. Newman, *Planetary Accounting*,
https://doi.org/10.1007/978-981-15-1443-2_1

environment. But the PBs were not designed to be used in this way. They were intended as Earth-system science indicators of the extent and urgency of the problem. They were not meant to be a guide to resolving it. Regional and national environmental targets have been developed using the PBs as guidance, and scientists have established frameworks to try and link the PBs to environmental accounting systems. However, each of these works has limitations, both in their connectivity to the PBs and in their applicability to environmental management. The PBs just do not relate simply to human activity or to different levels of government.

Herein lies the basis of this book. We (the authors) have developed a framework that sets out how to break up global environmental problems into more manageable chunks that you—as an individual, a chief executive, a city councillor, or a national committee member—can tackle. We call it "Planetary Accounting" because it is about creating a series of budgets that can be used to help us manage the global environment from whatever scale of influence we hold. Planetary Accounting can be used to quantify what we all need to do in order to return to and live within the safe operating space of the Planetary Boundaries.

There were three key discoveries that led to the eventual solution of Planetary Accounting (which will be described in much greater detail in the book). The first discovery came from management theory. Current environmental management practices are often through top-down governance or private management. These practices are based on out-of-date theories of environmental management such as the tragedy of the commons—the concept that, in the absence of enforced rules, humans are unable to share resources (such as forests or fisheries) without overusing and exploiting them. More recent findings in the fields of behaviour change, commons management, and change theory can be used to show that a more effective approach would be one that applies at different scales and in different ways—from individual bottom-up initiatives, to business and private sector efforts, to traditional top-down governance. Importantly, these approaches should be coordinated by a general system of rules that has the flexibility to accommodate these different centres of activity. We call this a "poly-scalar approach".

The second key discovery was a critical insight from accounting theory that only if there are standards or limits can you create serious change; that "you cannot manage what you do not measure". Humans have become very good at estimating past, present, and even future environmental impacts from human activity. The ongoing measurement and monitoring of environmental assets (e.g. forests, land, or fisheries) and of our impacts on these is called environmental accounting. It is now common for businesses, cities, and nations to keep environmental accounts—i.e. to track environmental impacts and the state of environmental assets over time against targets and benchmarks. Environmental accounting is an important tool for helping humans to reduce our impacts on the planet as it can be used to inform decision-making at any scale of activity. However, the key limitation is that for most environmental impacts, there is no clearly determined limit. Targets for managing impacts are typically set based on percentage reductions from the status quo or on industry best practice. Such targets are arbitrary. Even where targets are based on local policy or legislation, these are not often based on scientific limits. Moreover, targets that focus on the (negative) status quo can seem exhausting. They convey a sense that people will always need to reduce, reduce, reduce. Rather than generating change, such targets can lead to a feeling of helplessness and despair in the face of our seemingly insurmountable environmental problems. There is a need to connect concrete, science-based targets with existing environmental accounting practices.

The third, and perhaps the most important, discovery that informed this research came from theories about environmental indicators, that different types of environmental indicators serve different purposes. The European Environment Agency developed a framework to categorize environmental indicators, the Driver, Pressure, State, Impact, Response (DPSIR) framework (EEA 2005). Any environmental indicator can be classified as either a Driving force (a human need, such as the need for fuel), a Pressure (a flow to the environment, such as CO_2 emissions), a State (describing the state of the environment, such as the concentration of CO_2 in the atmosphere), an Impact (describing a change in State, such as global warming), or a Response (describing a human response to the environment, e.g. the Paris Agreement).

Human activity directly influences Drivers and Pressures, but only indirectly influences States and Impacts. Both types of indicators are important, but they serve different purposes. State and Impact indicators communicate the status quo. Pressure and Driver indicators communicate action. Confusingly, the PB indicators are not all in the same DPSIR category. Three of the PB indicators are Pressures, five are States, and one is an Impact. The PBs given in State and Impact indicators give an overview of planetary health. They communicate information such as that the concentration of CO_2 in the atmosphere and the rate of species extinctions are too high. This is an important message, but does not quantify what needs to be done. In contrast, the PBs given in Pressure indicators are limits for human activity. They do communicate what we should do, for example, how much freshwater we can safely consume or how much nitrogen we can fixate without pushing the global environment to a point of irreversible change.

The insight that the PB indicators are of varying DPSIR categories is not new. There have been several projects in which the DPSIR framework was used to make the PBs relevant to policy. The first attempt to translate the PBs into policy was the development of national targets for Sweden (Nykvist et al. 2013). The authors identified the need to translate the PBs into Pressure indicators. However, they only achieved this for one of the PBs, the PB for CO_2 concentration. They did not translate the remaining State and Impact PB indicators into limits for Sweden. A later adaptation of the PBs to targets for Switzerland also references the variance in the PB indicator classifications. However, the indicators selected for Switzerland in this study also vary between DPSIR categories. No one has previously translated the full set of PBs into Pressure-based indicators.

The three discoveries described here led us to the conclusion that what was needed was a new type of environmental accounting that incorporated Pressure indicators and science-based targets while retaining enough flexibility to be applied in a ploy-scalar way.

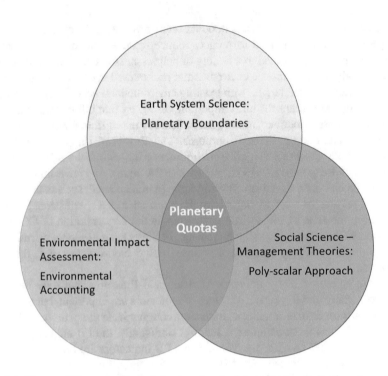

Fig. 1.1 The novel Planetary Quotas bring together the latest advances in Earth-system science (the Planetary Boundaries), environmental impact assessment (environmental accounting), and the social science of management theory (poly-scalar management)

This book thus introduces a new set of global limits, based on the Planetary Boundaries, but using Pressure indicators, the Planetary Quotas (PQs). The PQs bring together the insights from the three fields of research described above by using the DPSIR framework to connect the Planetary Boundaries and environmental accounting in a way that can be used at different scales and for different types of human activity (see Fig. 1.1).

The PQs create the foundation for our new concept—Planetary Accounting— comparing the results of environmental impact assessments to global scientifically determined limits (see Fig. 1.2).

It is our intention that Planetary Accounting will provide a platform for the behaviour, policy, organizational, and technological change that will be needed if we are to return to and live within the safe operating space.

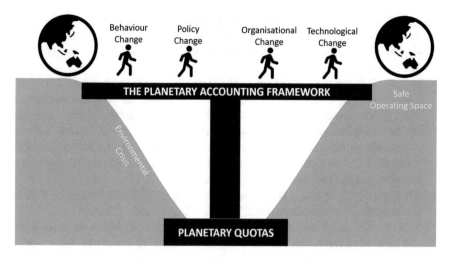

Fig. 1.2 The Planetary Quotas form the foundations of the Planetary Accounting Framework—a platform for change

1.2   A Guide to This Book

This book is divided into three sections. The first provides a review of the literature that led us to the discoveries described in this introduction and goes on to show how these helped us to arrive at the solution of a Planetary Accounting Framework and the Planetary Quotas.

The second section describes the research methods used in the development of the overall concept as well as giving some background to and the scientific basis for each of the Quotas.

The last section of the book summarizes the Planetary Quotas, describes Planetary Accounting and shows how this can be used, and discusses the strengths, limitations, potential applications, and future work in this area.

Section 1

- Chapter 2: The Science of Anthropogenic Climate Change
  This book is based on the assumption that human activity can change and is changing the state of the planet. In recognition that not everyone believes that this assumption is true, this chapter presents the scientific evidence that supports this theory and addresses the key arguments made against it.
- Chapter 3: The Holocene, the Anthropocene, and the Planetary Boundaries
  This chapter provides an overview of past attempts to define the limits for human impacts on the planet. It highlights some of the challenges people have faced in determining global limits and shows why we believe that the Planetary Boundary framework is the most robust and transparent definition for global limits at this point in time and thus forms the basis of Planetary Accounting.
- Chapter 4: Managing the Earth System: Why We Need a Poly-scalar Approach

This chapter summarizes some of the past and present theories of how best to manage shared environmental resources and of how to generate change. It concludes with the case that a poly-scalar approach is needed to manage the Earth system.

- Chapter 5: Environmental Accounting, Absolute Limits, and Systemic Change
  This chapter gives a historical account of environmental impact assessment methods and environmental accounting practices. It shows that a key limitation of most environmental accounting is that the impacts of human activity determined through environmental impact assessments cannot be compared to scientific limits. It demonstrates that absolute, scientific limits are likely to be a key component to achieving systemic change.
- Chapter 6: Resolving the Disconnect Between Earth-System Science, Environmental Management Theory, and Environmental Accounting
  This chapter shows how the DPSIR framework from environmental accounting can be used to show why the Planetary Boundaries are not accessible or applicable to human activity. It concludes that key constraint for using the Planetary Boundaries is that they are not all of a single DPSIR category and that in order to make them accessible, they should be translated into a uniform set of Pressure category indicators.

Section 2

- Chapter 7: Translating the Planetary Boundaries into Planetary Quotas
  This chapter provides a high-level introduction to the Planetary Boundaries and Planetary Quotas. It then describes the overall methodology for the project and the methods used to translate the PBs into PQs.
- Chapter 8: A Planetary Quota for Carbon Dioxide
  This chapter presents the background and need for a Planetary Quota for carbon dioxide and shows the detailed methodology used to derive this.
- Chapter 9: A Planetary Quota for Methane and Nitrous Oxide
  This chapter presents the background and need for a Planetary Quota for methane and nitrous oxide and shows the detailed methodology used to derive this.
- Chapter 10: A Planetary Quota for Forestland
  This chapter presents the background and need for a Planetary Quota for reforestation and shows the detailed methodology used to derive this.
- Chapter 11: A Planetary Quota for Ozone-Depleting Substances
  This chapter presents the background and need for a Planetary Quota for ozone-depleting substances and shows the detailed methodology used to derive this.
- Chapter 12: A Planetary Quota for Aerosols
  This chapter presents the background and need for a Planetary Quota for aerosols and shows the detailed methodology used to derive this.
- Chapter 13: A Planetary Quota for Water
  This chapter presents the background and need for a Planetary Quota for water and shows the detailed methodology used to derive this.
- Chapter 14: A Planetary Quota for Nitrogen

This chapter presents the background and need for a Planetary Quota for nitrogen and shows the detailed methodology used to derive this.

- Chapter 15: A Planetary Quota for Phosphorus
  This chapter presents the background and need for a Planetary Quota for phosphorus and shows the detailed methodology used to derive this.
- Chapter 16: A Planetary Quota for Biodiversity
  This chapter presents the background and need for a Planetary Quota for biodiversity and shows the detailed methodology used to derive this.
- Chapter 17: A Planetary Quota for Imperishable Waste

  This chapter presents the background and need for a Planetary Quota for imperishable waste and shows the detailed methodology used to derive this.

  Section 3

- Chapter 18: The Planetary Accounting Framework
  This chapter brings the Planetary Quotas together as a suite of global limits for human activity and shows how these can be used by outlining a high-level Planetary Accounting Framework to inform many different aspects of human activity. It outlines the key strengths and weaknesses of the research project and identifies areas of future work.

1.3 Publications

A summary of this book has been published in the new Springer-Nature-BMD journal *Sustainable Earth:*

Meyer, K. and Newman, P. 2018. Planetary accounting – a quota-based approach to managing the Earth system. *Sustainable Earth* 1:4–25.

The work described in this book was undertaken under the umbrella of a PhD by Kate Meyer and supervised by co-author Peter Newman. The thesis, which provides an extended description of much of the technical information presented in this book, is available from the Curtin University Library (Meyer 2018).

References

EEA (2005) EEA core set of indicators. Guide. Office for Official Publications of the European Communities, Luxembourg

Meyer K (2018) Planetary quotas and the planetary accounting framework - comparing human activity to global environmental limits. PhD, Curtin

Nykvist B, Persson Å, Moberg F, Persson LM, Cornell SE, Rockström J (2013) National environmental performance on planetary boundaries: a study for the Swedish Environmental Protection Agency. Sweden, Swedish Environmental Protection Agency

Rockström J, Steffen W, Noone K, Persson A, Chapin FS, Lambin E, Lenton TM, Scheffer M, Folke C, Schellnhuber HJ, Nykvist B, De Wit CA, Hughes T, Van Der Leeuw S, Rodhe H, Sorlin S, Snyder PK, Costanza R, Svedin U, Falkenmark M, Karlberg L, Corell RW, Fabry VJ, Hansen J, Walker B, Liverman D, Richardson K, Crutzen P, Foley J (2009) Planetary boundaries: exploring the safe operating space for humanity. Ecol Soc 14:32
Steffen W, Richardson K, Rockström J, Cornell SE, Fetzer I, Bennett EM, Biggs R, Carpenter SR, De Vries W, De Wit CA, Folke C, Gerten D, Heinke J, Mace GM, Persson LM, Ramanathan V, Reyers B, Sörlin S (2015) Planetary boundaries: guiding human development on a changing planet. Science 347:1259855

Part I
Earth-System Science, Management Theory, and Environmental Accounting: The Disconnect

Chapter 2
The Science of Anthropogenic Climate Change

We have to wake up to the fierce urgency of the now
Jim Yon Kim

Abstract There is scientific evidence to suggest that human activity, in particular the release of greenhouse gases to the atmosphere, is currently causing the climate to warm up. There is only a 1 in 100,000 chances it is not human activity. A warmer climate is predicted to be unfavourable for humanity. However, there are some people who dispute the theory of anthropogenic climate change, arguing either that the climate is not changing, that the change is not caused by human activity, or that it is not important.

Data taken from multiple sources shows a clear warming trend since preindustrial times. 2016 was the hottest year on record and July, 2019 was the hottest month on record. Average temperatures are now more than 1 °C higher than during preindustrial times.

Greenhouse gases trap shortwave radiation and therefore heat into the atmosphere. Approximately equal quantities of greenhouse gases are emitted and absorbed naturally every year. Human activity releases a relatively small amount of greenhouse gases to the environment compared to natural processes. However, we absorb very little of what we emit. Thus, there is a net flow of these gases to the environment from human activity.

There are no natural factors which correlate with the current warming trends. The primary theory for natural warming is that it is caused by changes in the solar cycle. However, the amount of energy coming from the Sun has been reducing since 1980, and warming has continued.

There have been higher levels of greenhouse gases in the atmosphere before and life has flourished. However, this has been during stable climate conditions. Rapid increases or decreases of greenhouse gases in the past have been highly destructive to life on Earth.

Human-induced climate change is expected to mean less favourable conditions for humankind, as we leave the "zone of comfort" within which cities and agriculture have been able to develop over the past 10,000 years.

© Springer Nature Singapore Pte Ltd. 2020
K. Meyer, P. Newman, *Planetary Accounting*,
https://doi.org/10.1007/978-981-15-1443-2_2

2.1 Introduction

The premise of this book is that human activity is changing the Earth system and that this is one of the greatest risks to humankind. Not all people share this view. The debate between climate change advocates and climate change sceptics in the media is prolific, heated, and emotive. There is name-calling; advocates and sceptics call one another "deniers" and "alarmists". Both sides frequently use the term "myths" to describe the arguments of those opposing their views. Much of the debate in the media has lost any connection with the science at question. Arguments often focus around *who said what* rather than scientific facts. This chapter is therefore setting out to show why Earth-system science is important by applying it to the climate change issue, and also why it is important to understand the human issues behind the controversies.

Despite almost unequivocal scientific evidence to support their case, many advocates continue to fall back on the feeble argument that 97% of climate scientists believe in anthropogenic global warming rather than citing scientific facts to support their case. The views of scientists do not constitute scientific evidence. Further, it is a misrepresented statistic. The paper behind this statistic (Cook et al. 2013) states that only 32.6% of all articles on climate change expressed any opinion on human-induced warming in the abstract. Of this 32.6%, 97% supported the theory. However, the remaining 67.4% of the articles analysed in the study did not articulate an opinion for or against human-induced warming in the abstract.

Both sides argue unethical conduct from the other side, driven by perceived conflicts of interest. Advocates often accuse sceptics of being funded by the fossil fuel industry or other financially interested parties. Sceptics argue that there is a conspiracy that scientists' claims of anthropogenic climate change are a bid to gain governmental control over energy consumption. One group of sceptics hacked email servers at a leading institution for climate research and posted (misrepresented) snippets of emails on the Internet to support their conspiracy theory. Neither side of the debate is innocent.

The stakes of the debate and therefore the emotions of those debating are high. From the advocates' point of view—the stakes are the wellbeing of the planet. They (like we) believe that failure to act will lead to severe consequences for humankind. However, to act is unlikely to be a minor undertaking. Sceptics are reluctant to make changes of the order of magnitude believed necessary by advocates based on what they believe to be uncertain science.

It is understandable therefore that the debate is so fierce and sensitive. However, at the core, it is a debate over scientific evidence. This chapter does not explore who said what or the motivations of sceptics or advocates. It presents the scientific evidence for and against anthropogenic (human caused) climate change and attempts to address both sides of the argument with transparency. The chapter begins by introducing the concept of the Earth system. It then presents the core evidence for anthropogenic climate change. Those who do not believe in anthropogenic climate change generally fall into one of three categories: those who do not believe the climate is changing; those who believe the climate is changing but do not believe that this is caused by human activity; and those who believe human activity is changing the climate but do not agree that this is important. This chapter goes on to address the key arguments made against anthropogenic climate change under these three categories.

2.2 Earth as a System

The sum of the planet's physical, chemical, and biological processes is known as the Earth system. Everything in the Earth system belongs to one of four subsystems or "spheres": the geosphere (land), hydrosphere (water), atmosphere (air), and biosphere (life). The four spheres are interconnected by Earth-system processes—such as evaporation, transpiration, and photosynthesis—that store, transfer, and transform matter and energy according to the laws of physics and chemistry (Skinner and Murck 2011). The relationships between the processes are complex. There are many feedback loops—responses to external influences which can dampen or amplify change. Feedback loops can lead to tipping points—points of rapid, runaway, irreversible change—even when the original perturbation is removed (Scheffer et al. 2001; Lenton et al. 2008). For example:

> People emit carbon to the atmosphere (an external influence). Carbon dioxide is a greenhouse gas which means is traps heat in the atmosphere. As the atmosphere heats up, this causes ice to melt leaving behind dark blue ocean. Ice is reflective. Dark blue ocean is not. This means that more heat can be absorbed by Earth's surface (assuming other factors affecting Earth's albedo do not change). More heat means more ice melts. Earth's reflectivity reduces further, and the feedback loop continues.

This is only one example. There are many other feedback loops that affect climate change and other Earth-system processes. Some, like the melting ice, are positively reinforcing, i.e. they accelerate change. Other feedback loops help to stabilize Earth-system processes. These are called negative feedback loops. The risk that we face today is that we may reach a tipping point—where the positive feedback loops accelerate change—resulting in rapid and possibly irreversible change, beyond our control. The state of the Earth system can change very rapidly. For example, the transition from the last glacial period, the Younger Dryas, to the current interglacial, is thought to have happened over only a few decades. In Greenland, temperature changes of as much as 10 °C per decade are believed to have occurred during this period (Severinghaus et al. 1998).

The Earth system can operate in many different states. Each state is typically separated by a relatively short period of rapid change. Average global temperature is not the only variable that changes from one environmental state to another. The chemical composition of the atmosphere, the amount of energy the Earth's surface receives from the Sun, the ratio of ocean to land area of Earth's surface, and the number and type of species inhabiting Earth are all examples of variables that can differ between different environmental states.

2.3   We Are at Risk of Changing the State of the Planet

Since the industrial revolution, human emissions of CO_2 have accelerated exponentially. Since humans evolved, CO_2 levels have been approximately 350 ppm. The atmosphere currently contains more than 400 parts of carbon dioxide (CO_2) per million parts of atmosphere (ppm). This is higher than any level measured since we began measuring the amount of carbon dioxide in the atmosphere. The rate of increase in CO_2 emissions correlates with the rate of increase in CO_2 concentration. Approximately half of all anthropogenic emissions remain in the atmosphere. There is *very high confidence*[1] that the amount of CO_2, and other greenhouse gases (GHGs) including methane, and nitrous oxide in the atmosphere is higher than it has been in 800,000 years (IPCC 2013). To put this into context, homo sapiens are believed to have evolved 300,000 years ago. Humans have never experienced such high concentrations of CO_2. More alarming to scientists than the high concentration of greenhouse gases is the rate of increase of these gases in the atmosphere. CO_2 levels are currently increasing at a rate between 100 and 200 times faster than the rate of increase that occurred at the end of the last Ice Age. There are other periods in history where CO_2 levels have increased rapidly—at rates of the same order of magnitude as rates of increase today. These past events have been highly destructive to life—causing mass global extinctions—i.e. more than 75% of the existing species on Earth went extinct in a short period of time.

It is not the CO_2 in the atmosphere per se that is concerning. Rather, concerns are for the anticipated impacts to the Earth system that this could cause. There has been a very strong correlation between CO_2 and global average temperatures over the last 800,000 years (McInnes 2014) (see Fig. 2.1). It is *virtually certain*[1] that globally, the troposphere has warmed since the mid-1900s (IPCC 2013). It is *extremely likely* that more than half the warming that has been recorded for average surface temperatures from 1950 to now occurred because of human activity (IPCC 2013).

It is not certain what increased average temperatures would mean for humanity, but predictions are not optimistic. It is *likely*[1] that increased temperatures will lead to global average increase in rainfall and that the rainfall distribution will change so that wet areas become wetter and dry areas become drier. It is *very likely* that the Atlantic Meridional Overturning Circulation (AMOC), the global flow of oceans that is an important component of Earth's climate system, will weaken, although it is *very unlikely* that it will collapse altogether this century. It is *very likely* that arctic sea ice will continue to shrink and thin and that global glacier volume will continue to decrease. It is *very likely* that sea levels will continue to rise and that the rate of sea level rise will increase. The likelihood of future increases to the frequency and/or intensity of extreme weather events ranges from *more likely than not* (for tropical

---

[1] The terms used to describe likelihood correspond to scientific probabilities as follows: "virtually certain"—>99% "extremely likely" >95%, "very likely" >90%, "likely" >66%, "more likely than not" >50%, and "very unlikely" <10%. The term "very high confidence" conveys a 9/10 chance of being correct.

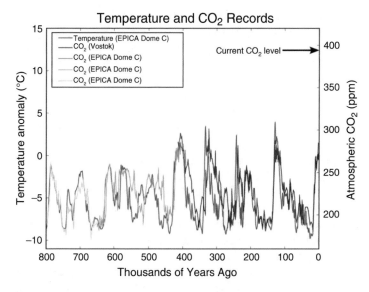

Fig. 2.1 Average change in temperature with respect to average Holocene temperatures and atmospheric concentration of CO_2. (McInnes 2014, CC BY-SA 3.0, CC BY-SA 3.0: Creative Commons License allows reuse with appropriate credit)

cyclone activity) to *virtual certainty* (for warmer days and nights over most land areas) (Stocker et al. 2013; IPCC 2013).

2.3.1   Understanding CO_2 and Temperature

Figure 2.1 shows 800,000 years of CO_2 and temperature data, yet we have only been recording CO_2 levels since 1950. The estimates of past CO_2 levels in the atmosphere before 1950 are based on measurements of ancient air that is stored in glacial ice. The ancient air samples can also be used to estimate past temperatures because the composition of air changes with changing temperatures. Other independent evidence is then used to support ice core data. For example, the distance between tree rings indicates tree growth rate which is influenced by both temperature and CO_2 levels. There is tree ring data spanning 10,000 years which, until recent years, showed a strong correlation with the ice-core data. Fossilized leaves can be used as another indication of past CO_2 levels. There is approximately 400,000 years of leaf fossil data, and this correlates closely with ice core data. Lake and ocean sediments change with temperature, rainfall, and snowfall and can also be used to support ice core data (NOAA 2018).

Ice layers accumulate over hundreds of thousands of years which protect ancient ice from melting, thus storing important information about the past climate. However, heat from bedrock below the ice slowly melts the oldest ice so that until recently, ice core data had only been found dating back 800,000 years.[2] To determine CO_2 levels and temperatures prior to 800,000 years ago, proxy data such as isotopes found in shells and fossils of ancient marine organisms have been used. This data provides insight into the climatic conditions for the entire Phanerozoic period—i.e. the geological eon beginning 540 million years ago that we are still in today, albeit at a low resolution (Veizer et al. 1999; Berner 1991; Berner and Kothavala 2001; Crowley and Berner 2001; Royer et al. 2004).

There have been several analyses of CO_2 and temperature data for the Phanerozoic period. Of these, most found a positive correlation between CO_2 and temperature, i.e. low CO_2 levels overlapped with extensive glaciations and high CO_2 levels did not (Berner 1991, Berner and Kothavala 2001, Crowley and Berner 2001, Royer et al. 2004). One study did not find a positive correlation between the two (Veizer et al. 2000). However, it was found later that the temperature proxy data used in this study had not been corrected for seawater pH. Once the data had been updated, the same positive correlation could be seen (Royer et al. 2004).

One of the arguments put forwards by sceptics is that the Phanerozoic CO_2 and temperature data are not coupled. The basis of this argument is that one period of glaciation that occurred during this period does not appear to correlate with low CO_2 levels. This glaciation, known as the late Ordovician glaciation, occurred approximately 440 million years ago. Some sceptics have used this data to suggest that this period of glaciation coincided with very high levels of CO_2 somewhere between 2400 and 9000 ppm (Berner and Kothavala 2001).

The correlation between CO_2 and temperature found for the Phanerozoic period pertains to *extensive* periods of glaciation only. This is not because the data shows that shorter period of glaciation occurred during high levels of CO_2. It is because the data is too coarse to draw any conclusions about shorter periods of glaciation or any short period at all. Extensive periods of glaciation do correlate with low levels of CO_2.

The CO_2 data available for this period is in intervals of approximately 10 million years (Royer et al. 2004). In contrast, the late Ordovician glaciation is thought to have lasted less than 1 million years (Royer et al. 2004). Only one datum exists close to the period of the glaciation (Royer 2006). This proxy data point cannot be more accurately dated than a point in time between 450 and 443 million years ago. It is conceivable that the CO_2 levels dropped during the Ordovician glaciation. There is geochemical evidence to support this theory. Carbon cycle modelling for this period suggests that CO_2 levels may have dropped to 3000 ppm during the glaciation (Kump et al. 1999).

Even at 3000 ppm it may seem unlikely that a period of glaciation could occur. Current CO_2 levels are a little over 400 ppm, and glaciers and arctic sea ice are melting. However, CO_2 is only one driver of average temperature. There are other factors

---

[2] Recently ancient ice, approximately 2.7 million years old, has been discovered. Voosen, P. 2017. Record-shattering 2.7-million-year-old ice core reveals start of the Ice Ages. *Science.* There is not yet enough ice to draw strong conclusions from the findings. However, scientists are hopeful that this discovery will lead to a greater understanding of ancient climatic conditions.

such as Earth's orbit, and the intensity of radiation from the Sun, that must also be considered. The current CO_2-ice threshold—the level of CO_2 below which glaciation is possible—is estimated to be 500 ppm. This means that if all other factors such as the Sun's radiation and Earth's orbit around the Sun remain constant, when the CO_2 levels reach 500 ppm, there will be no more ice on Earth. During the Late Ordovician the solar constant was 4% less than it is now, i.e. 4% less solar energy was entering Earth's atmosphere. Royer (2006) estimated that in that case, the CO_2-ice threshold would be approximately 3000 ppm. This means that glaciation could occur at any CO_2 concentration below 3000 ppm. Their estimate is consistent with other estimates for the CO_2-ice threshold during the Late Ordovician period which range from 2240 to 3920 ppm (Crowley and Berner 2001; Crowley and Baum 1995; Gibbs et al. 1997; Kump et al. 1999; Herrmann et al. 2004). The high CO_2 levels shown during the Late Ordovician period do not negate the positive correlation found between CO_2 and temperature during the Phanerozoic period.

There is substantial evidence that CO_2 and temperature have correlated in the past.

2.3.1.1 Correlation or Causation

Of course, correlation does not mean causation (see Box 2.1). However, there is also scientific evidence that there is a causal relationship between CO_2 and temperature. CO_2 is a greenhouse gas (GHG). This means that it traps infrared radiation from the Sun in the atmosphere and warms the planet. Without GHGs, Earth would not be habitable. With too many GHGs, Earth would become too hot, and Earth would also not be habitable. One of the arguments by sceptics is that the correlation of CO_2 concentration and temperature doesn't necessarily indicate causation. Further, they argue that historically, CO_2 has followed temperature—i.e. that temperature increases have caused CO_2 levels to increase rather than vice versa.

The relationship between CO_2 and temperature is more complex than this argument suggests. One does not lead the other. There are many positive and negative feedback loops that relate the two variables. For example, at the end of the last Ice Age, Earth's orbital cycle led to warming in the Arctic. This warming caused large amounts of ice to melt—reducing the salinity of local sea water. The freshwater influx altered the natural ocean cycles and led to warming of the Southern Hemisphere oceans. The warmer oceans could not hold as much CO_2, so large amounts of CO_2 were released into the atmosphere. The increase in CO_2 in the atmosphere trapped more solar radiation and therefore increased the temperatures leading to more melting of ice and release of CO_2. In this example temperature increase was the initial driver that set the changes in motion. However, after this initial change, CO_2 then drove temperature increase (Shakun et al. 2012).

In summary, it is *extremely likely* that the CO_2 released by humans has caused and will continue to cause the global average temperature to increase (IPCC 2013).

Box 2.1 Spurious Correlations
There is a website dedicated to finding correlations in data sets which are clearly unrelated. For example, they show a 95.86% correlation between the per capita consumption of mozzarella cheese and civil engineering doctorates awarded from 2000 to 2009 (Vigen 2018) (Fig. 2.2).

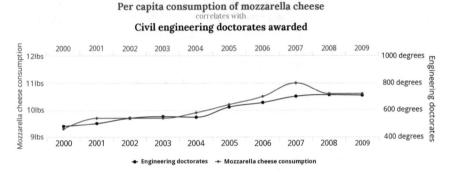

Fig. 2.2 Data correlation between cheese consumption and doctorates awarded highlights that correlation does not mean causation. (Vigen 2018, CC BY 4.0)

2.4 Is the Earth Getting Warmer?

The global average temperature has not risen to the same extent as CO_2 levels since the industrial revolution. However, there is a definite warming trend since the beginning of the upward trend in CO_2 levels (see Fig. 2.3). Data from many different sources shows that temperatures have risen by approximately 1.1 °C since the industrial revolution (WMO 2017; Met Office 2018; NOAA 2017; Climate Copernicus 2017; NASA 2017).

In addition to the temperature data, there are many other lines of evidence that support the theory that the planet is warming. Greenland and Arctic ice sheets are getting smaller (Kjeldsen et al. 2015). Glaciers are melting (Kjeldsen et al. 2015). Sea levels have risen between 3 and 9 in. (Cole 2017).

There are five key arguments frequently made as evidence that the planet is not warming:

1. In the 1970s scientists predicted that we were heading for an Ice Age.
2. 1934 was the warmest year on record.
3. There has been no warming since 1998.
4. The warming recorded reflects only the urban heat island effect.
5. Antarctic sea ice is increasing.

These theories are discussed below.

2.4.1   1970s Predictions of an Imminent Ice Age

One of the most widely touted arguments against the theory that Earth's climate is warming is that in the 1970s climate scientists were predicting cooling. From 1940 to the early 1970s, there was a cooling trend (see Fig. 2.4). In the early 1970s, there was a period when the cooling trend appeared to be accelerating. In 1971 and again in 1972, there was an abrupt increase in the snow coverage area in the northern hemisphere of 12%. Areas that were usually void of snow in summer stayed covered

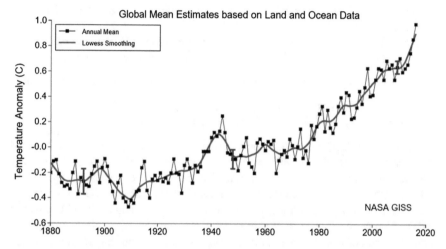

Fig. 2.3 Global average temperatures since the industrial revolution showing a clear upward trend. (CCO 1.0 BY SA 3.0, CCO 1.0: Universal Public Domain Dedication)

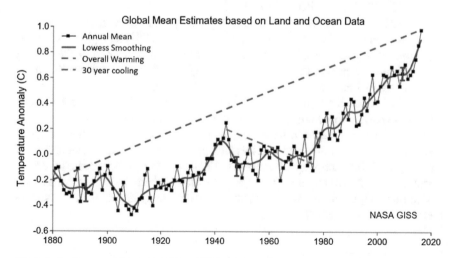

Fig. 2.4 Cooling trend from the 1940s to 1970s within the overall warming trend since the industrial revolution. (Adapted Image CCO 1.0, CCO 1.0: Universal Public Domain Dedication)

all year round, and this continued in subsequent years. There were extreme droughts in Africa. Growing periods in England decreased by 2 weeks from 1950 to 1970, resulting in losses in the order of 100,000 tons of grain per year. Climatologist Kenneth Hare predicted that if 1972 conditions persisted for more than 3 years, the world's population would not be able to be sustained (Lahaye and Hindson 1996).

During this period, several scientists did predict that the cooling trend would continue and warned of an imminent Ice Age:

- Rasool and Schneider (1971) predicted that the cooling effects of aerosols would outweigh the warming effects of CO_2.
- Bryson (1974) also concluded that aerosols cooling effects would exceed the effects of CO_2.
- Barrett and Gast (1971) suggested that warming was unlikely on the basis of his predictions that CO_2 doubling in the atmosphere would take 340 years.

All of these papers predicting global cooling made two important assumptions:

1. Human emissions of carbon dioxide will have a warming effect.
2. Human activity can influence global average temperatures.

The idea that CO_2 in the atmosphere could warm up the planet was introduced well before this period in 1896 by Svante Arrhenius (1896). In the 1930s, Guy Callendar was cautioning that warming was underway. By the 1970s, the theory that CO_2 and other greenhouse gases had a warming effect was not in dispute. By this time, however, it was understood that human activity, including the burning of fossil fuels (which were very dirty at the time) could also have a cooling effect through the production of aerosols.[3] Many scientists were working to understand the complex interactions between human-induced cooling and warming, e.g. Rasool and Schneider (1971), Bryson (1974), Kellogg and Schneider (1974), Manabe and Wetherald (1975), and Mitchell (1972).

In the 1970s, it was thought that an interglacial period (a warm period between Ice Ages) could not last more than approximately 10,000 years (this theory has since been disproven). It is a little over 10,000 years since the end of the last Ice Age, so it is unsurprising that some scientists believed that a 30-year cooling trend was indicative of the end of the current interglacial period.

Despite the cooling trend, and the timing of the last Ice Age, many scientists in the 1970s were still predicting warming. Schneider, who in 1971 predicted cooling, co-authored another paper in which the authors acknowledged that the effects of aerosols were poorly understood. In this later paper, the authors estimated 0.5 °C of warming by 2000 (Kellogg and Schneider 1974). Another author predicted 0.8 °C warming over the twentieth century (Manabe and Wetherald 1975). Another suggested that CO_2 would be more influential in its warming effects than aerosols in their cooling effects (Mitchell 1972).

---

[3] Aerosols are small particles suspended in the atmosphere which absorb and scatter light. See Chap. 12 for a more detailed description of aerosols and their effects.

The idea of an imminent Ice Age sparked much attention at the time. Hundreds of articles were published on the subject across a wide range of media. Some current-day articles still cite these media articles as evidence of "alarmism" by scientists, e.g. Newman (2017). Some sceptics draw from this previous, incorrect theory to conclude that there is no reason to believe scientists who are now warning of global warming.

There are certainly similarities in the nature and message of the articles from the 1970s and media articles on global warming today.

- A 1975 Newsweek article called "A Cooling World" begins:

  There are ominous signs that the earth's weather patterns have begun to change drastically
  ....

  The article warns of pending food shortages and extreme weather conditions—a "grim reality".

- An earlier article from *Time* magazine published in 1974 called "Another Ice Age?" which predicts cooling of 2.7 °F (1.5 °C) makes similarly ugly predictions:

  Whatever the cause of the cooling trend, its effects could be extremely serious if not catastrophic ....

- A 1973 Science Digest article—Brace Yourself for Another Ice Age—warns:

  the end of the present interglacial period is due soon.

- A more recent article published in the Washington Times in 1998 by Fred Singer, an atmospheric physicist at George Mason University, gained wide traction against the theory of global warming. His article refers to a report by the US National Academy of Sciences (NAS 1975):

  ... But this exaggerated concern about global warming contrasts sharply with an earlier NAS/NRC report ... There, in 1975, the NAS "experts" exhibited the same hysterical fears – this time, however, asserting a "finite possibility that a serious worldwide cooling could befall the Earth within the next 100 years ....

It is understandable that these quotes, taken in isolation, might lead a person to feel that it was all unfounded hysteria in the 1970s and therefore conclude that the situation is similar now. However, the details of the articles cited above tell quite a different story to the quotes on their own. For example, A Cooling World goes on to say that the causes of Ice Ages remain "a mystery" and quotes a different section of the same NAS report cited by Singer which reads: "Our knowledge of the mechanisms of climate change is at least as fragmentary as our data .... Not only are the basic scientific questions largely unanswered, but in many cases we do not know enough to ask the key questions". Another Ice Age? qualifies the prediction for cooling as "at best an estimate". This article highlights that some scientists believe that the cooling trend is only temporary and, importantly, that "all agree that more information is needed about the major influence on the earth's climate". Brace Yourself for Another Ice Age clarifies that "soon" (the timeframe identified for the end of the current interglacial period) referred to geologically soon—i.e. anything from 200 to

2000 years and states that "scientists seem to think that a little more carbon dioxide in the atmosphere could warm things up a good deal". Climate Change: Chilling Possibilities states "the cooling trend observed since 1940 is real enough ... but not enough is known about the underlying causes to justify any sort of extrapolation" and "by the turn of the century, enough carbon dioxide will have been put into the atmosphere to raise the temperature of earth half a degree [C]".

Singer's (1998) article grossly misrepresents the NAS report (1975) which is neither hysterical nor certain in its findings. The forward reads:

> ... we do not have a good quantitative understanding of our climate machine and what determines its course. Without the fundamental understanding, it does not seem possible to predict climate ....

The report is a call for a major research programme on the climate on the basis of a growing awareness of the reliance of humanity's economic and social stability on the climate and on the potential for human activities to influence it. Singer's quote that the experts in the report are asserting a "finite possibility that serious worldwide cooling could befall the Earth within the next 100 years" comes from this paragraph:

> ... there seems little doubt that the present period of unusual warmth will eventually give way to a time of colder climate, but there is no consensus as to the magnitude or rapidity of the transition. The onset of this climatic decline could be several thousand years in the future, although there is a finite probability[4] that a serious worldwide cooling could befall the earth within the next 100 years. The question remains unresolved. If the end of the interglacial is episodic in character, we are moving toward a rather sudden climatic change of unknown timing, although as each 100 years passes, we have perhaps a 5% greater chance of encountering its onset. If, on the other hand, these changes are more sinusoidal in character, then the climate should decline gradually over a period of thousands of years. These climatic projections, however, could be replaced by quite different future climatic scenarios due to man's inadvertent interference with the otherwise natural variation.

The full paragraph shows careful consideration of the possible future climatic conditions as understood, as well as a high transparency regarding the lack of knowledge at the time. It is not hysterical. Nor does it suggest any certainty regarding the theory of an imminent Ice Age.

In 1976, the weather returned to normal and the cooling trend was abruptly over. It is now thought that two main drivers led to the temporary cooling:

1. A surge in the emissions of aerosols after World War II from the burning of dirty fossil fuels (i.e. the aerosols released led to cooling)
2. A cool phase in the Pacific Ocean Cycle (this cool phase has again masked warming over the past two decades—this is discussed later in this chapter)

---

[4] Note that Singer used the word "possibility" in his quote where actually the word was "probability"—thus altering the sentence from an acknowledgement that there *could* be imminent cooling to a suggestion that this was *likely*.

The argument by sceptics that the 1940–1970s cooling trend and scientists' predictions of a possible Ice Age should not be taken as evidence against global warming because:

1. Warming trends have been observed over a much longer period—since 1750
   (WMO 2017; Met Office 2018; NOAA 2017; Climate Copernicus 2017;
   NASA 2017).
2. There is strong scientific evidence that emissions of CO_2 cause temperature
   increase—this was not debated in the 1970s concerns over global cooling. Nor is
   it debated today.
3. One of the theories thought to have caused the temporary cooling is the emission
   of atmospheric aerosols. Scientists still believe that aerosols have a cooling
   effect on the atmosphere. Moreover, many are concerned that as we reduce aerosol emissions to improve air quality, some of the masking effects of aerosols on
   global warming will diminish and warming will accelerate more rapidly.

2.4.2   1934: The Warmest Year on Record

Another point commonly presented as evidence against global warming is that 1934
was the warmest year on record. This is not true. 1934 was the warmest year in the
United States (US). It was not the warmest year globally. However, the argument
gained a lot of traction with sceptics because an error in the GISS data had previously shown 1998 to be the hottest year in the United States. This error has been
taken as evidence that recent warming may not be as high as the data suggests. It is
also used as evidence that the temperature data cannot be trusted.

2012 is now the hottest recorded year in the United States. The hottest year
globally was 2016 (WMO 2017; Met Office 2018; NOAA 2017; Climate
Copernicus 2017; NASA 2017). It is plausible that there may be more errors in the
data that have not yet been discovered. However, the global temperature data is a
compilation of many different data sources. It is unlikely that isolated errors such
as this will affect the global trends to a noteworthy degree. When the mistake was
found in the US data, this only had a 0.185 °C/decade impact to the global data—
i.e. the change to the global mean was less than 1/1000 of a degree.

2.4.3   The 1998–2015 Warming Hiatus

In 2014 American politician Ted Cruz stated in a CNN interview that:

> The last 15 years, there has been no recorded warming. Contrary to all the theories that they
> are expounding, there should have been warming over the last 15 years. It hasn't
> happened.

Box 2.2 Understanding Temperature Data

Average temperature data for the past 150 years is predominantly taken from a compilation of measurements made at sea. Sailors and ship captains have always been interested in sea and air temperatures and many logged extensive measurements.

The sailors did not anticipate that this data would be used in the future to understand global conditions. As such, there were no consistent methods to test the temperatures. Methods also changed over time. In the early days, methods were coarse—sailors would drop a bucket overboard and measure the temperature in the bucket. Later, water temperatures were measured automatically when water was pumped into the engine room. The different methods result in slight variations in the measured temperature and actual temperature.

Scientists therefore use various mechanisms to interpret the data to attempt to account for the potential variations between measuring methods.

This statement was based on atmospheric temperature data, which, at the time, did show a hiatus in warming from 1998 until 2005. Data pertaining to natural systems is almost always noisy and non-linear with unexpected outliers and anomalies. Past and future global warming is about long-term warming trends. This does not mean that every year should be warmer than the last. Nor does it mean there will not be short periods of constant or reducing temperatures. There was a $\approx$30-year cooling period from the 1940s to 1970s within the last $\approx$250 years of warming (see Sect. 2.4.1). A 15-year hiatus in warming would not necessarily constitute evidence against the theory that the long-term trend is warming.

The perceived pause did puzzle scientists for many years however. Many developed theories to explain it, but none were conclusive. During this time, many sceptics took the combination of the pause, and lack of robust explanation to be strong evidence that the climate was not changing. In 2015, scientists at NOAA realized that the data interpretation methods they had used for ocean data was overestimating early temperatures and underestimating more recent temperatures (see Box 2.2). When they updated the method of interpretation to account for these inconsistencies, the data no longer showed a pause in warming.

Sceptics used this revision of data as evidence of a conspiracy theory. However, an independent study reviewed raw data from buoys, against satellite and sensor data to assess the NOAA findings (Hausfather et al. 2017). Their results matched the amended NOAA results.

2.4.4   Global Warming or the Urban Heat Island Effect

Box 2.3 The Urban Heat Island Effect
Urban areas have dark, heat absorbing surfaces. They have less evapotranspiration (release of water to the atmosphere by plants) than non-urban areas.
They often have poor air flow due to high-rise buildings. These factors contribute a localized temperature increase known as the urban heat island effect.
Urban areas can be as much as 2 °C hotter than the surrounds. Even within
cities, there is much local variation between temperatures recorded over roads
and temperatures recorded over vegetation and light-coloured infrastructure.

Some argue that the warming trends shown in the data are not showing average
global temperature increases but rather localized increases in urban temperatures
caused by the urban heat island effect (UHI) (see Box 2.3). Most temperature sensors are located in urban areas. The premise is that the urban locations of temperature sensors results in falsely high measurements. One paper suggests that as much
as half of the global warming trend recorded from 1980 to 2002 can be attributed to
the UHI effect (McKitrick and Michaels 2007).

The concerns that the UHI effect could skew climate data are shared by climate
change advocates. NASA and GISS go to considerable efforts to account for potential impacts of the UHI effect in their data. To do this, they compare long-term
trends of cities to long-term trends in nearby rural areas and then adjust the urban
trends accordingly so that the data is not skewed. The impacts of the UHI effect on
the data prior to these adjustments have thus far been found to be minor.

2.4.5   Antarctic Sea Ice Is Increasing

Southern sea ice is increasing. How can this be if the climate is getting warmer?
This paradox is often used as evidence against climate change. However, this argument is not compelling. There is data that shows that atmospheric temperatures are
increasing. There are also data that show that the oceans, including the Southern
Ocean, are warming. In fact, the Southern Ocean, surrounding the Antarctic sea ice,
is getting warmer by approximately 0.17 °C per decade. This is faster than the
global ocean warming trends of 0.1 °C.

Localized increases in ice in the face of global average temperature increases in
the oceans, atmosphere, and Earth's surface do not constitute evidence that the
global climate is not warming. The Earth system is complex with global and local
climates and climate phenomena. Nonetheless, there are theories to explain that
increasing sea ice may be caused by global warming (see Box 2.4).

Box 2.4 Theories on Why Antarctic Sea Ice Is Increasing
There are two main reasons believed to be causing the increase in sea ice. The first is the effects of the ozone hole. Lower stratospheric ozone in this region has strengthened cyclonic winds that move sea ice around creating polynyas—areas of open water. An increased number of polynyas mean more sea ice (Gillett and Thompson 2003).

The second reason is increased precipitation caused by the warmer atmosphere. Increased snowfall onto land increases sea ice. Increased snow and rain falling on the surrounding water reduce the salinity of the water. Normally the ocean currents bring deep warm water to the region, which rises and melts sea ice. Reduced salinity in oceans increases stratification (the separation between shallow cold water and deep warmer water or vice versa). The increased stratification means that less warm water rises to the surface and therefore less of the sea ice is melted.

2.5 Human Activity Is the Main Driver for the Changes to the State of the Earth

Some sceptics accept that the climate is getting warmer but debate that human activity is causing this. Scientific investigation does not result in certain proof of a hypothesis. Rather, a hypothesis is proposed, and evidence is gathered to either support or dispel this. As such, it is not possible to prove without doubt that the human emissions of CO_2 (among other things) are causing global average temperatures to increase. However, we can examine the evidence to support this, and the evidence against it. We can also consider other hypotheses and the evidence around these to draw conclusions as to the most likely theory.

The evidence to support the hypothesis of human-induced warming includes:

- CO_2 and other greenhouse gases have a warming effect.
- CO_2 concentration and temperature have been very closely linked for the last 800,000 years—the period during which we have a lot of data. The data spanning the last 540 million years also shows strong correlation between these variables.
- Human emissions of CO_2 since the industrial revolution have increased substantially.
- The concentration of CO_2 in the atmosphere has increased since the industrial revolution at a rate that has not occurred in the last 800,000 years.
- The amount of CO_2 released naturally into the environment is approximately equal to the amount of CO_2 absorbed by the environment (so one would expect that the release of additional CO_2 into the environment by humans would alter the balance).

The IPCC have concluded their assessment of climate science with a detailed probability analysis that shows the chances of recent climate change not being caused by human activity are around 1 in 100,000.

The main hypothesis proposed as an alternative cause of global warming is that it is that it is caused by changes in the solar cycle. The evidence cited against anthropogenic warming is that human emissions of CO_2 are insignificant.

2.5.1 Is Climate Change Caused by the Sun?

Total solar irradiance (the amount of energy coming from the Sun) is higher now than it was in 1750. In the past, solar irradiance and global temperature were closely coupled, suggesting that the Sun was the main driver of temperature change until recently. However, since the 1980s, temperature increases have accelerated, while solar irradiance has been dropping (see Fig. 2.5). This decoupling of solar irradiance and global temperatures is further evidence to support the hypothesis that human activity is the main cause of global warming.

2.5.2 Human Emissions of CO_2 Are Insignificant

Without human intervention, approximately 750 billion tons of CO_2 are emitted naturally every year. Humans emit approximately 34 billion tons of CO_2 each year, only a fraction of natural emissions. This means that human emissions constitute less than 4.5% of natural emissions. However, without human intervention,

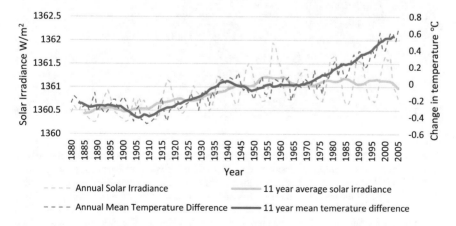

Fig. 2.5 Average solar irradiance and average global temperatures have not correlated since the 1980s. (Solar irradiance data from Kopp 2015, temperature data from Schmidt 2018)

approximately 750 billion tons of CO_2 are absorbed naturally out of the atmosphere. The balance is the 34 billion tons of CO_2 emitted by humans.

The carbon cycle is self-regulating to a point. If the concentration of CO_2 in the atmosphere increases, the oceans, land, and vegetation absorb a little more carbon. This natural regulation has absorbed approximately half of anthropogenic CO_2 emissions (Ciais et al. 2013). However, the other half, ≈ 17 $GtCO_2$ each year, remains in the atmosphere.

There are approximately 3000 $GtCO_2$ in the atmosphere. This means that each year human activity is adding CO_2 in the order of magnitude of 0.5% of the total amount. However, since 1870, the end of the industrial revolution, humans have emitted approximately 2000 $GtCO_2$. Approximately half of this has been absorbed into the carbon cycle.[5] This means that we have added approximately 1000 $GtCO_2$ to the atmosphere, in approximately 150 years. This is an increase of 30%. At current rates, we will have increased the amount of CO_2 in the atmosphere by 50% by mid-century. In context, human emissions of CO_2 do not appear so insignificant (Box 2.5).

Box 2.5 Altering a Balanced System: The Bathtub Analogy for Climate Change

Consider a bathtub that has just the right amount of water in it. If you pull out the plug, the water will begin to flow out. However, if you turn on a tap so that the amount of water running into the bath is equal to the amount of water escaping the bath, the water level will remain constant. This is (more or less) how the natural carbon cycle works. Now consider that every minute, someone adds a few drops of water to the bath, just 0.5% of the total amount of water in the bath (representing human emissions of CO_2 to the atmosphere). The effect of additional water is barely noticeable at first. However, in less than 2 h, more than half of the original volume of water would have been added. Even if the bath was only half full to begin with, it would only be a matter of time, before it overflowed. The atmosphere will not overflow. However, the analogy helps to demonstrate the potential impacts of very small inputs to a balanced cycle.

---

[5] The amount of CO_2 that can is absorbed naturally is not fixed but rather corresponds to the amount of CO_2 emitted. For example, emissions between 1980 and 1989 were much lower, at approximately 25 $GtCO_2$/year. During this period the oceans and the land only absorbed approximately 13 $GtCO_2$/year over the background rates. Once again, a little over half of the anthropogenic emissions were absorbed. It is not known whether the natural carbon cycle will continue to absorb half of anthropogenic emissions. Moreover, the absorption of additional CO_2 into the natural carbon cycle is not without consequence. For example, increased CO_2 absorption by the oceans is the cause of increasing ocean acidity which is a dangerous consequence for marine life.

2.6   Climate Change Matters

Some sceptics argue that there is no reason to be concerned about the changing climate because the climate has always changed. They make the point that life on Earth has flourished in periods of high CO_2 such as the Eocene and the Cretaceous periods. It is true that the climate has always changed. Figure 2.6 shows average temperatures over the past 500 million years. Temperatures have been as much as 14 °C hotter and several degrees cooler than current temperatures.

During the past periods of high CO_2, at least those during which life flourished, the greenhouse gases were in balance. There are other periods in history where the CO_2 levels have increased rapidly—as we are increasing them today. These events have been destructive to life—and are thought to be the cause of some of the past global mass extinctions when almost all life on Earth went extinct.

Humans have only been around for approximately 300,000 years. During this time the climate has changed a lot. Humans have survived through several Ice Ages and an interglacial that is warmer than today (see Fig. 2.6). However, for much of human history, humans subsisted as hunter-gatherers. It is only in the last 10,000 years, under the much more stable conditions that humans have developed into settled cities and agricultural societies. The "zone of comfort" that was the Holocene had no greater band of temperature than 1 °C, and now we are leaving that band and heading into completely new territory where there is considerably greater variations in climate with significantly more atmospheric energy in the storms and extreme temperatures.

Even during the last 10,000 years, there have been brief (several hundred year) periods of cooler and warmer temperatures including the Late Antique Little Ice Age from 536 to 660 AD, the Little Ice Age 1300–1700 AD, and the Early Medieval Warm Period from 950 to 1200 (see Fig. 2.7). These periods of warmer and cooler temperatures were not as ubiquitous as recent warming (Stocker et al. 2013). Nonetheless, these periods are marked by great social upheaval. The end of the Late Antique Ice Age led to large-scale migrations that contributed to the decline of the Western Roman Empire. The Mayan collapse can be linked to draughts caused by climate change. Even without global climate change, past societies have collapsed because of human impacts on the environment. This has happened even after warnings that collapse was imminent. Cities have been deserted after failure of their inhabitants to heed cautions of overconsumption of natural resources (Diamond 2005).

The climate has always changed, but there have been negative consequences for life during times of rapid change. The difference now is that this is the first time that humans are causing the change. It is also the first time that humans have the capacity to prevent the change. Global limits may not yet have been exceeded to the point of no return, but evidence suggests that the point of no return may be close. Exceeding environmental limits is not a theoretical concern. It is a real one.

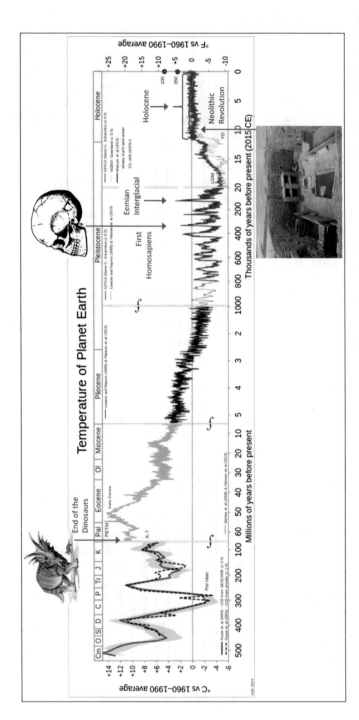

Fig. 2.6 Average global temperatures on Earth during different geological epochs over the last 500 million years. (Adapted from Fergus 2014, Benito 2006, and McKay 2014, CC BY-SA 3.0, CC BY-SA 3.0: Creative Commons License allows reuse with appropriate credit)

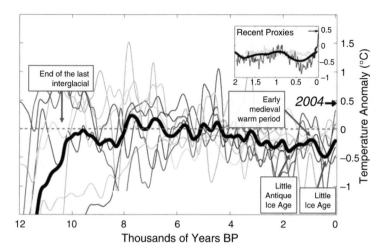

Fig. 2.7 Temperature variations during the Holocene. (Adapted from Rohde 2010, CC BY 3.0, CC BY-SA 3.0: Creative Commons License allows reuse with appropriate credit). The coloured lines show proxy data, black line shows the average of these datasets. The dotted line shows mid-twentieth-century average temperature

2.7 Conclusion

The scientific evidence shows with near certainty that climate change is happening, that it is predominantly caused by human activity, and that the potential implications are grim. Thus, a major planetary limit does appear to have already been exceeded and suggests we must take seriously all the others as well as climate change. Climate change is only one Earth-system process. There are many others, for example, the nitrogen and water cycles, that are also being altered by human activity. It is important to understand what level of impact from humans the Earth system can withstand before there are major changes to the Earth-system function so that we can manage our activity accordingly.

The following chapter provides an overview of past and current attempts to define global environmental limits within which human activity should be managed and introduces the Planetary Boundaries—environmental limits that together define a safe operating space for humanity.

References

Arrhenius S (1896) On the influence of carbonic acid in the air upon the temperature of the ground. Lond Edin Dub Philos Mag J Sci 41:237–275

Barrett E, Gast PF (1971) Climate change. Science 173:982–983

Benito JM (2006) Qafzeh skull-11.png. Wikimedia

Berner RA (1991) A model for atmospheric CO2 over Phanerozoic time. Am J Sci 291:339–376

Berner RA, Kothavala Z (2001) Geocarb III: a revised model of atmospheric CO2 over phanerozoic time. Am J Sci 301:182–204

Bryson RA (1974) A perspective on climatic change. Science 184:753–760

Ciais P, Sabine C, Bala G, Bopp L, Brovkin V, Canadell J, Chhabra A, Defries R, Galloway J, Heimann M, Jones C, Le Quéré C, Myneni RB, Piao S, Thornton P (2013) Carbon and other biogeochemical cycles. In: Stocker TF, Qin D, Plattner G-K, Tignor M, Allen SK, Boschung J, Nauels A, Xia Y, Bex V, Midgley PM (eds) Climate change 2013: the physical science basis. Contribution of working group I to the fifth assessment report of the intergovernmental panel on climate change. Cambridge University Press, Cambridge; New York, NY

Climate Copernicus (2017) Earth on the edge: record breaking 2016 was close to 1.5C warming. EU Climate Copernicus. European Centre for Medium-Range Weather Forecasts, Reading

Cole S (2017) NASA science zeros in on ocean rise: how much? How soon? NASA, Washington, DC

Cook J, Nuccitelli D, Green SA, Richardson M, Winkler B, Painting R, Way R, Jacobs P, Skuce A (2013) Quantifying the consensus on anthropogenic global warming in the scientific literature. Environ Res Lett 8:024024

Crowley TJ, Baum SK (1995) Reconciling late Ordovician (440 Ma) glaciation with very high (14X) CO2 levels. J Geophys Res Atmos 100:1093–1101

Crowley T, Berner R (2001) Enhanced - CO2 and climate change. Science 292:870–872

Diamond JM (2005) Collapse: how societies choose to fail or succeed. Viking; Allen Lane, New York, NY; London

Fergus G (2014) All Palaeotemps.png. Wikimedia

Gibbs MT, Barron EJ, Kump LR (1997) An atmospheric pCO2 threshold for glaciation in the late Ordovician. Geology 25:447–450

Gillett NP, Thompson DWJ (2003) Simulation of recent southern hemisphere climate change. Science (New York, NY) 302:273

Hausfather Z, Cowtan K, Clarke DC, Jacobs P, Richardson M, Rohde R (2017) Assessing recent warming using instrumentally homogeneous sea surface temperature records. Sci Adv 3:e1601207

Herrmann AD, Patzkowsky ME, Pollard D (2004) The impact of paleogeography, pCO2, poleward ocean heat transport and sea level change on global cooling during the Late Ordovician. Palaeogeogr Palaeoclimatol Palaeoecol 206:59–74

IPCC (2013) Summary for policymakers. In: Stocker TF, Qin D, Plattner G-K, Tignor M, Allen SK, Boschung J, Nauels A, Xia Y, Bex V, Midgley PM (eds) Climate change 2013: the physical science basis. Contribution of working group I to the fifth assessment report of the intergovernmental panel on climate change. Cambridge University Press, Cambridge; New York, NY

Kellogg WW, Schneider SH (1974) Climate stabilization: for better or for worse? Science 186:1163–1172

Kjeldsen KK, Korsgaard NJ, Bjørk AA, Khan SA, Box JE, Funder S, Larsen NK, Bamber JL, Colgan W, Van Den Broeke M, Siggaard-Andersen M-L, Nuth C, Schomacker A, Andresen CS, Willerslev E, Kjær KH (2015) Spatial and temporal distribution of mass loss from the Greenland Ice Sheet since AD 1900. Nature 528:396

Kopp G (2015) Total solar irradiance data. University of Colorado, Boulder, CO

Kump LR, Arthur MA, Patzkowsky ME, Gibbs MT, Pinkus DS, Sheehan PM (1999) A weathering hypothesis for glaciation at high atmospheric pCO2 during the Late Ordovician. Palaeogeogr Palaeoclimatol Palaeoecol 152:173–187

Lahaye T, Hindson E (1996) Global warming. Harvest House Publishers, Eugene, OR

Lenton TM, Held H, Kriegler E, Hall JW, Lucht W, Rahmstorf S, Schellnhuber HJ (2008) Tipping elements in the Earth's climate system. Proc Natl Acad Sci U S A 105:1786–1793

Manabe S, Wetherald RT (1975) The effects of doubling the CO2 Concentration on the climate of a general circulation model. J Atmos Sci 32:3–15

McInnes L (2014) CO2-temperature-records.svg. Wikimedia

McKay S (2014) Skara Brae - geograph.org.uk - 1446822.jpg. Wikimedia

McKitrick RR, Michaels PJ (2007) Quantifying the influence of anthropogenic surface processes and inhomogeneities on gridded global climate data. J Geophys Res Atmos 112:D24S09

Met Office (2018) 2016: one of the warmest two years on record. UK Met Office, Devon

Mitchell JM (1972) The natural breakdown of the present interglacial and its possible intervention by human activities. Quatern Res 2:436–445

NAS (1975) Understanding climate change: a program for action. National Academy of Sciences, Washington, DC

NASA (2017) NASA, NOAA data show 2016 warmest year on record globally. NASA, Washington, DC

Newman A (2017) Climate alarmists have been wrong about virtually everything. The New American

NOAA (2017) State of the climate: global climate report for December 2016

NOAA (2018) Past climate. National Oceanic & Atmospheric Administration - Earth System Research Laboratory - Global Monitoring Division. Available: https://www.climate.gov/maps-data/primer/past-climate. Accessed 3 Nov 2018

Rasool SI, Schneider SH (1971) Atmospheric carbon dioxide and aerosols: effects of large increases on global climate. Science 173:138–141

Rohde R (2010) Holocene temperature variations. Wikimedia

Royer DL (2006) CO2-forced climate thresholds during the Phanerozoic. Geochim Cosmochim Acta 70:5665–5675

Royer DL, Berner RA, Montanez IP, Tabor NJ, Beerling DJ (2004) CO2 as a primary driver of Phanerozoic climate. GSA Today 14:4–10

Scheffer M, Carpenter SR, Foley JA, Folke C, Walker BH (2001) Catastrophic shifts in ecosystems. Nature 413:591–596

Schmidt G (2018) Global land-ocean temperature index. NASA, Washington, DC

Severinghaus JP, Sowers T, Brook EJ, Alley RB, Bender ML (1998) Timing of abrupt climate change at the end of the Younger Dryas interval from thermally fractionated gases in polar ice. Nature 391:141

Shakun J, Clark P, He F, Marcott SA, Mix AC, Liu Z, Otto-Bliesner B, Schmittner A, Bard E (2012) Global warming preceded by increasing carbon dioxide concentrations during the last deglaciation. Nature 484:49

Singer F (1998) Scientists add to heat over global warming. Washington Times. 5 May 1998

Skinner BJ, Murck B (2011) The blue planet: an introduction to earth system science. Wiley, Hoboken, NJ

Stocker TF, Qin D, Plattner G-K, Alexander LV, Allen SK, Bindoff NL, Bréon F-M, Church JA, Cubasch U, Emori S, Forster P, Friedlingstein P, Gillett N, Gregory JM, Hartmann DL, Jansen E, Kirtman B, Knutti R, Krishna Kumar K, Lemke P, Marotzke J, Masson-Delmotte V, Meehl GA, Mokhov II, Piao S, Ramaswamy V, Randall D, Rhein M, Rojas M, Sabine C, Shindell D, Talley LD, Vaughan DG, Xie S-P (2013) Technical summary. In: Stocker TF, Qin D, Plattner G-K, Tignor M, Allen SK, Boschung J, Nauels A, Xia Y, Bex V, Midgley PM (eds) Climate change 2013: the physical science basis. Contribution of working group I to the fifth assessment report of the intergovernmental panel on climate change. Cambridge University Press, Cambridge; New York, NY

Veizer J, Ala D, Azmy K, Bruckschen P, Buhl D, Bruhn F, Carden GAF, Diener A, Ebneth S, Godderis Y, Jasper T, Korte C, Pawellek F, Podlaha OG, Strauss H (1999) 87Sr/86Sr, δ13C and δ18O evolution of Phanerozoic seawater. Chem Geol 161:59–88

Veizer J, Godderis Y, François LM (2000) Evidence for decoupling of atmospheric CO2 and global climate during the Phanerozoic eon. Nature 408:698

Vigen T (2018) Spurious correlations. Available: http://www.tylervigen.com/spurious-correlations. Accessed 17 Apr 2018

Voosen P (2017) Record-shattering 2.7-million-year-old ice core reveals start of the ice ages. Science

WMO (2017) WMO confirms 2016 as hottest year on record, about 1.1C above pre-industrial era. World Meteorological Organisation, Geneva

Chapter 3
The Holocene, the Anthropocene, and the Planetary Boundaries

There is no planet B
Richard Branson

Abstract People have been trying to determine environmental limits for the planet since as early as the 1600s. However, this task is inherently difficult as it requires a high level of value judgement. Assumptions regarding lifestyle, technology, and population underpin most past attempts to determine planetary limits.

The Holocene is the period of time that started 11,650 years ago. This is only a small fraction of human history which can be traced back 300,000 years. Prior to the Holocene, the climate was highly variable. Humans lived as hunter-gatherers moving from place to place to survive. The Holocene was an unusually stable and warm period in human history. In this nurturing environment, humans developed from hunter-gatherers to urban and agricultural settled societies. The Holocene is the only state in which we know humanity can thrive with anything like the 7.5 billion humans being supported today.

We have now left the Holocene and are in the transition to the Anthropocene. This new geological epoch was named to acknowledge human influence on the state of the planet. The state of the planet in the Anthropocene is not yet determined, but at current trends in human activity, predictions are for a much hotter and less stable climate, a "hot-house Earth" scenario.

In 2009, the Planetary Boundaries were proposed. These are environmental limits for the planet within which the climate and other environmental conditions in the Anthropocene are likely to resemble those of the Holocene. There are no assumptions regarding lifestyle, technology, or population underpinning the Planetary Boundaries. The limits are based on the latest scientific understanding of the planet's environmental processes. At least four of the Planetary Boundaries have been exceeded.

It would be prudent for humans to try to return to and operate within the Planetary Boundaries so that the risk is low of changing the state of the planet from a Holocene-like state which is favourable to humanity and especially to the kind of civilization based on cities and agriculture, to one where substantial collapse of the population is likely.

© Springer Nature Singapore Pte Ltd. 2020
K. Meyer, P. Newman, *Planetary Accounting*,
https://doi.org/10.1007/978-981-15-1443-2_3

3.1 Introduction

The task of defining the planet's environmental limits is not straightforward. There are no biophysical laws which define the limits. It is reasonable to assume that even with several degrees of global temperature increase, Earth would continue to spin on its axis. When people refer to planetary limits, they are not usually referring to limits for the planet per se. Rather, what is normally meant by the term planetary limits is "limits for maximum planetary change that is acceptable for humanity". This means that there is a level of value judgement inherent in any definition of planetary limits. An acceptable level of planetary change is likely to be different for different people. For some, the only acceptable conditions might be those in which humanity is thriving. For others, it might be acceptable for humans to be simply surviving. Environmental limits for the planet would vary according to these different definitions of acceptable conditions (see Fig. 3.1).

This chapter begins with an overview of how people have defined planetary limits in the past and the limitations of these definitions. The Holocene and Anthropocene, the past and future states of the planet, are then introduced to give context to the Planetary Boundaries. The chapter concludes with the case that humanity should aim to live within the Planetary Boundaries and an explanation of what these are.

3.2 Planetary Limits: A Brief History

The idea of planetary limits can be traced back to as early as the 1600s. Dutch scientist, Antonie van Leeuwenhoek, estimated Earth's "carrying capacity", the maximum human population Earth could support as 13.4 billion (F.N.L.P 1962). His calculation was based on his estimate for the maximum population of Holland, multiplied by his estimate of the ratio of global inhabited land area to the area of Holland (F.N.L.P 1962).

There have been at least 94 estimates of carrying capacity since Leeuwenhoek's (Cohen 1995). Normally, with increasing numbers of studies of a scientific phenomenon, one would expect convergence of results over time. Interestingly, rather than trending towards a single value, the range in estimates of carrying capacity has increased over time (Cohen 1995). Current estimates range from <1 billion to >1000 billion with one outlying estimate at 1 sextillion (10^{20}) (Cohen 1995). Some of these estimates for the maximum number of humans Earth can support are substantially less than today's population. Some might argue that they must be incorrect on this basis. However, of the 7.5 billion people alive today, almost half live below the poverty line, without basic needs being met. This suggests that we are already living beyond the planet's capacity to equitably support all of us, in the long term, whether this also means that we are beyond Earth's carrying capacity or not.

To estimate how many people the Earth can support, one must first make some assumptions about what sort of lifestyle those people should have. These assump-

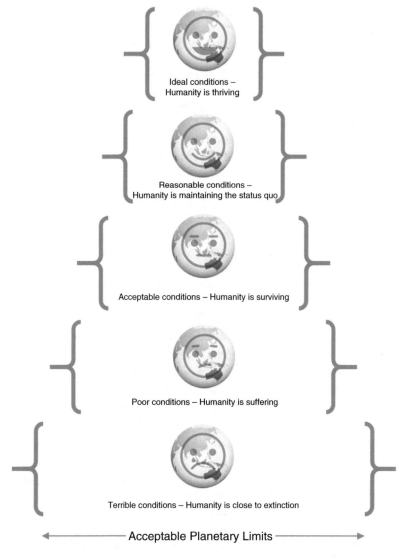

Fig. 3.1 Planetary limits are dependent on the conditions deemed acceptable for humanity

tions are the basis of the high variation in the results. Estimates based on a high consumption lifestyle such as those experienced by many of the world's wealthiest today will be relatively low. In contrast, estimates which assume that only basic needs for food, water, and shelter must be met will be relatively high. The estimation of 1 sextillion includes cannibalism as a means to nourish the population (Franck et al. 2011).

The concept that human consumption could exceed the planet's capacity to provide to us became widespread when it was bought to light in Malthus' seminal

Essay on the Principle of Population (1798). Malthus postulated that food supply would be unable to keep up with population growth because food supply had linear growth and the population was increasing exponentially. Later, as agriculture became mechanized and efficiencies improved, food production increased exponentially, and his theory seemed to be without foundation. However, in the late 1960s and early 1970s, as concerns mounted regarding limited oil supplies, other authors picked up the notion of the planet's limits (Ehrlich 1971; Meadows et al. 1972; Nordhaus and Tobin 1971).

The book *Limits to Growth* (Meadows et al. 1972) is the earliest recorded attempt at defining global scale limits. The authors do not specify global limits per se; rather, they discuss findings of a computer simulation of global future scenarios. The scenarios are modelled against five basic factors they deemed to be the key determinants in limiting growth: population increase, agricultural production, nonrenewable resource depletion, industrial output, and pollution.

Since the 1970s the idea of planetary limits, or more broadly, *sustainability* has become increasingly popular. The primary Oxford definition for the word *sustainable* is "able to be maintained at a certain rate or level" (Oxford Dictionaries 2014). The use of this term to refer to operating within planetary limits seems to have developed gradually. The earliest use of the term in the context of planetary limits may have been by Meadows et al. (1972) when they wrote:

> It is possible to alter these growth trends and to establish a condition of ecological and economic stability that is sustainable far into the future.

The idea of ecological sustainability is now so popular that the secondary Oxford definition for *sustainable* is "conserving an ecological balance by avoiding depletion of natural resources". The term sustainable development can be traced back to The Brundtland Commission (1987) which defines this as:

> development that meets the needs of the present without compromising the ability of future generations to meet their own needs.

The Brundtland definition of sustainability is still broadly used, but in their book and in many following global conferences, sustainable development became focussed on how development could achieve environmental goals and social goals along with the normal economic goals. This became known as *triple bottom line* sustainability. This concept was documented further by Spreckley (1987) and stems from the term "bottom line" as it is used in accounting—the total profit or loss—recorded at the end of a financial statement.

Another approach to defining sustainability or planetary limits called The Natural Step has been to develop a general set of rules for use in decision-making, typically prohibiting:

- Increasing concentrations of substances from Earth's crust in the atmosphere, hydrosphere, and biosphere
- Increasing concentrations of substances produced by society
- Excessive physical manipulation or over-harvesting

with a fourth rule pertaining to societal wellbeing (Robèrt et al. 2013; Broman et al. 2000; Goodland and Daly 1996; Daly 1990; Azar et al. 1996).

In 1989 Eugene Odum published a book entitled *Ecology and our Endangered Life-Support Systems* (Odum 1989). It was an advanced book for its time as it addressed ecology as an integrated system with a focus on the importance of the entire biosphere, down to the smallest of ecosystems. Odum refers to the failed engineering of Apollo 13, which exploded just moments into its first flight. He postulates that if we cannot even get a relatively simple life-support system that we designed and built right, we should be wary of tampering with the far more complex life-support system of our planet.

A decade later, Vitousek et al. (1997) wrote a paper outlining human's "domination" of Earth's ecosystems. Their paper identifies the key impacts of humanity on the Earth system which broadly encompass the same key Earth-system processes that were later identified as having critical Planetary Boundaries (see Sect. 3.5).

Since the 1990s there have been many attempts at quantifying environmental limits for the planet. Some key examples are described below:

Ecological Footprint/Biocapacity—the term Ecological Footprint, first mentioned in academic literature in 1992 (Rees 1992), is perhaps the most famous system that includes reference to environmental limits. It is a measure of human impacts, the "Ecological Footprint", which can be compared to the ability of Earth to provide, the "biocapacity". Global biocapacity is the total biologically productive land on Earth. It is a function of both size and productivity and thus changes from year to year. It is currently approximately 12 billion global hectares (Global Footprint Network 2012; Hoekstra 2009; Vale and Vale 2013). If the Ecological Footprint is larger than the biocapacity, this tells us we are operating beyond the limits. Both the footprint and the biocapacity are assessed in terms of a weighted land area called a "global hectare" (Wackernagel and Rees 1996). The Ecological Footprint was not intended to be a stand-alone indicator of sustainability (Ewing et al. 2010). It was designed to capture impacts which compete for space (Galli et al. 2014). The results of Ecological Footprint assessments are often communicated as the point in the year that you have used your annual "budget" of biocapacity and/or the number of planets that would be needed if everyone acted in the same way as the subject of assessment (see Fig. 3.2).

This communication is very accessible and effective. It is likely one of the key reasons the Ecological Footprint has become such a popular metric. The problem lies in how it is calculated (explained later) and in the implication that a global Ecological Footprint equal to the global biocapacity, a footprint of "one Earth", is the planet's limit. The problem with using biocapacity as a global limit is that this is the total resource bank of biological material for all species—not just humans. There is no sublimit for human appropriation of the total biocapacity. This is not a sustainable state. Humans rely heavily on ecosystem services for survival. Further, critical ecosystems such as the Amazon rainforest will be necessary if we are to remain in a Holocene-like state.

Fig. 3.2   Kate Meyer's Ecological Footprint. (Credit: Global Footprint Network 2019)

There is no consensus in the literature as to an appropriate "biodiversity buffer"—the amount of biocapacity that should be retained for the maintenance of the Earth system. Galli et al. (2014) propose the use of EF as an indicator for biodiversity but do not suggest a limit. Numerous studies have proposed a "biodiversity buffer", i.e. the amount of biocapacity which should be retained for biosphere integrity as part of the EF. The estimates range dramatically from 1% (Fahrig 2001), 10–25% (Wackernagel et al. 2002; The Brundtland Commission 1987), 75% (Margules et al. 1988), to 99% (Fahrig 2001). The limited examples where real case studies have been used to derive figures for a biodiversity buffer align at a value of around 45% (Soulé and Sanjayan 1998; Margules et al. 1988). It should be noted that both studies found better results from a greater buffer, but at this level each species or community was represented more than once, a description approximately comparable to a maximum extinction rate.

Further, the amalgamation of environmental impacts into a single indicator, while effective for communication purposes, it reduces the utility of the framework. It tells us that we are over the limits, but there is not enough resolution in the system for it to help us to understand what to do to address this. Ecological Footprint is thus severely limited as a system with which to understand the planet's limits.

Planetary Guardrails (WBGU 1995)—The concept of guardrails was first developed in response to human-induced climate change. Guardrails in this context are defined as "quantitatively definable damage thresholds whose transgression either today or in future would have such intolerable consequences that even large-scale benefits in other areas could not compensate these" (WBGU 2011). Guardrails have been proposed for:

- Global warming—mean global temperature rise <2 °C from preindustrial times (WBGU 1995)
- Soil degradation—rate of soil erosion <1 tons/hectare/year (WBGU 2005)
- Protected, conservation areas >20% of global area of terrestrial and river ecosystems (WBGU 2001)
- Ocean acidification—pH decline compared to preindustrial pH < 0.2 units (WBGU 2006)
- Long-lived and harmful anthropogenic substances (WBGU 2014)
- The loss of phosphoros (WBGU 2014)

These proposed limits are useful. However, the gradual addition of guardrails gives the sense of an incomplete picture. There is no evidence of a systematic review that ensures all critical limits are captured.

Tolerable Windows—This concept builds on the Planetary Guardrail system—using these as an upper limit for human impacts. This system advances the guardrails, by including minimum societal needs that must also be met. The "tolerable window" is where societal needs are met within the environmental limits (the guardrails) (Petschel-Held et al. 1999). This framework is limited by the limitations of the guardrail system.

Critical Natural Capital—This concept is based on the idea that environmental systems perform irreplaceable functions (Ekins et al. 2003), for example, bees pollinating plants. Critical natural capital is the level of environmental functions that are critical for humanity.

Each of these systems provides insightful contributions to the discussion of planet limits. However, none clearly define maximum acceptable planetary change for humanity.

3.2.1   The Population-Technology-Lifestyle Nexus

Ehrlich's popular book *The Population Bomb* (1971) began a long debate between environmentalists and economists regarding the importance of impacts of population growth on the environment versus the economy. There are also debates between environmentalists regarding the ethics of population control and the role of technology versus behaviour in solving our environmental crises.

Ehrlich and Holdren (1971) proposed a simplified mathematical equation to define the relationship between human impacts on the environment and population, affluence, and technology:

$$I = PAT$$

I = environmental impact which can be expressed in any unit of impact
P = population measured in persons
A = affluence measured in Gross Domestic Product (GDP) per person
T = technology in impact per unit of GDP

The premise of this equation is that an increase in population or affluence will have a proportional increase in impact. Likewise, any improvement in technology (a reduction in impact per GDP) will lead to a proportional decrease in impact. This is perhaps a useful tool in some instances. However, it makes incorrect assumptions about the simplicity of these three variables with respect to their impacts (Alcott 2010). Take, for example, an increase in a city's population by 100%. Using the IPAT equation we would estimate that the impacts of this city should double if technology does not change. However, the formula ignores the interconnectivity of the variables. An increase in population may lead to changes in both affluence and technology, for example, more efficient public transportation or recycling which generally increase with the size and density of a city, factors which would not have been considered by the IPAT equation.

Notwithstanding the above, the IPAT equation can be used to understand the underlying problem inherent in many of the past attempts to quantify planetary limits: there are value judgements and assumptions regarding either population, affluence (or lifestyle), or technology in many of these frameworks. The example above of the estimate for carrying capacity, where cannibalism is assumed to be reasonable sustenance, is an extreme example. On the other hand, Malthus's predictions that food supply would not match population growth (1798) probably seemed quite reasonable to most people based on the knowledge available at the time. Yet, they did not come true because Malthus assumed future food production would follow past trends and continue arithmetic growth. He did not predict the advancements in food production technology which enabled geometric growth that kept up with population growth rates.

Even definitions of limits that do include assumptions regarding technology or lifestyle have a level of value judgement as to the level of planetary change humans are willing to accept. The fundamental problem is that there are no biophysical laws which can be used to determine the limits (Van Vuuren et al. 2016).

3.3   The Holocene Epoch

The Holocene is the period of time which began 11,650 years before present (taken as the year 2000) (Severinghaus et al. 1998). Since the start of the Holocene, the state of the Earth system has been unusually stable, with average global temperature ranges of only ±1 °C. Homo sapiens evolved approximately 300,000 years ago, during the previous, Pleistocene epoch (Ewen 2017). The Pleistocene was a less stable epoch than the Holocene, marked by abrupt temperature changes as can be seen in Fig. 2.6. It is evident that homo sapiens can survive in different Earth-system states. Humankind survived through two Ice Ages and a brief interglacial period much warmer than current average temperatures (Jouzel and Masson-Delmotte 2007). However, during this period, for more than 280,000 years, humans subsisted as hunter-gatherers, moving from place to place so that they could survive.

The Holocene is both warmer and more stable than any other 10,000 year period of human history (IPCC 2007) (see Fig. 2.6). At the beginning of the Holocene, almost simultaneously, agriculture began in seven to eight geographically separate regions across the world (Bocquet-Appel 2011). This period is known as the Neolithic Revolution (Bocquet-Appel 2011).

Historians do not suggest that the change in climate was a driver for the civilization of humanity. They believe that humankind already had the intelligence and knowledge needed to begin the transition and the Holocene presented the needed "window of opportunity" for this to happen (Cook 2005). The warm and stable temperatures in the Holocene epoch enabled the rapid development of humans from hunter-gatherers to urban, agricultural, and industrial settled societies (Rockström et al. 2009b; Bocquet-Appel 2011).

The state of the planet during the Holocene—henceforth referred to as a Holocene-like state—is the only environmental state of the planet in which we know settled societies can thrive (Rockström et al. 2009b). Yet many scientists believe that the Holocene is over. They believe that we are in the transition to a new epoch—the Anthropocene (Rockström et al. 2009b; Crutzen 2002; Zalasiewicz et al. 2011). It is unknown whether society can thrive in other environmental states. It is also unknown what state the Anthropocene will have.

3.4   The Anthropocene Epoch

The *Encyclopedia of Global Environmental Change* (Trenberth 2002) lists key external forces which can alter the Earth system as:

- The Sun and its output
- The rate of Earth's rotation
- Sun-Earth geometry and the changing orbit of Earth around the Sun
- Earth's physical makeup:

  - Distribution of land and ocean
  - Geographic features on land
  - Ocean bottom topography and basin configurations
  - Mass and basic composition of the atmosphere and ocean

Many scientists now believe that human activity should be added to this list as we have become a primary driver of the Earth system (Steffen et al. 2005; Zalasiewicz et al. 2011; Rockström et al. 2009b). The new Anthropocene epoch is named as such to acknowledge the role humans are now thought to play in determining the state of the Earth system (Paul 2002; Crutzen 2002; Zalasiewicz et al. 2011).

Epochs are delineated through geochronology—rock dating. By definition, the beginning of any new epoch must be marked by a globally dispersed signal found in rock layers and deposits (Zalasiewicz et al. 2011). There is substantial evidence of such rock deposits which could be used to justify the start of the Anthropocene. Signals human activity has left in the rocks include radionuclides from nuclear testing, unburned carbon spheres from power stations, plastic pollution, aluminium and concrete particles, and residue from fertilizers (Lewis and Maslin 2015). It is interesting to note that many of these signals do not relate to climate change, further evidence of the multitude of global environmental impacts humans are having on the planet.

There are external factors which could change the state of the planet that are beyond human control, for example, the output of the Sun or the shape of Earth's orbit around the Sun (Trenberth 2002). In the 1970s, scientists believed that the Holocene was nearing its natural end and that without human intervention the Earth system would be headed into another Ice Age (Kukla et al. 1972). However, more recent evidence suggests that without human interference, the Holocene would be expected to continue for another several thousand or even tens of thousands of years (Berger and Loutre 2002). This estimation is based on scientist's understanding of the solar cycle (the changing output of the Sun) and patterns of change to Earth's orbit.

Of course, it is possible that the Holocene would come to a natural end sooner without human interference. There could be natural drivers we cannot predict, for example, a meteor that alters the orbit or the composition of the atmosphere or a major tectonic event (shifting plates under Earth's crust). However, such events are beyond our control. What we can influence are human impacts on the state of the Earth system.

Scientists are debating the precise start date of the Anthropocene. However, the dates proposed all fall within a timeframe of 1–2 centuries. This is an extremely short window in geological timeframes. Whatever date is finally agreed upon, we are currently operating at the intersection of the Holocene and the Anthropocene. The state of the planet is no longer truly a Holocene-like state (see Fig. 3.3). However, the long-term state of the Earth system in the Anthropocene is yet to be determined. The Anthropocene could mean a human-managed Holocene-like state, or an entirely new, warmer, unknown but likely unfavourable future (Rockström 2010), a "Hothouse Earth" (Steffen et al. 2018). A warmer Anthropocene is unlikely to occur through gradual, linear change (IPCC 2013b). Predictions are for dramatic and potentially irreversible change: substantial loss of species, devastating storms, significant sea level rise, and considerable displacement of communities (IPCC 2013a). It seems prudent thus that humans should aim for the Anthropocene to resemble the Holocene.

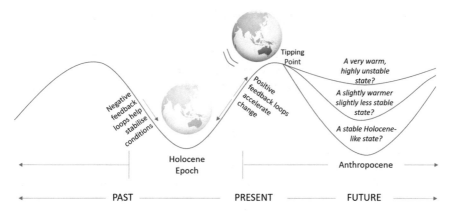

Fig. 3.3 We have left the safety of the Holocene epoch, but the state of the Anthropocene is still to be determined

3.5   Planetary Limits for a Holocene-Like State: The Planetary Boundaries

In 2009 Rockström et al. (2009a) proposed a new set of planetary limits, the Planetary Boundaries (PBs). This was updated in 2015 (Steffen et al. 2015). The Planetary Boundaries are global limits for Earth-system processes within which the risk of departing from a Holocene-like state is low.

The Planetary Boundary approach represents a breakthrough in defining planetary limits because the underlying assumption is that we ought to try to maintain a Holocene-like state. This changes the parameters of the problem as in this case there are biophysical laws which can be used to determine the limits. Further, the Planetary Boundaries are, for the most part, a measure of the state of the global environment. This differs to past attempts at defining limits which have attempted to measure human relationships with the Earth system. While there is value judgement in the determination of the point at which the limits should be, these key differences mean that there are no assumptions regarding lifestyle, technology, or population. The authors identified key Earth-system processes that, if altered too far by human activity, could lead to a change in Earth-system state away from a "safe" Holocene-like state (Rockström et al. 2009b). They assigned levels for each process based on the scientific literature. Where the science on particular limits was uncertain, the authors adopted the "precautionary principal"—i.e. in the absence of scientific consensus, the limits were set at the point at which the authors consider risk to be low based on the available scientific evidence. This decision framework in the development of the PBs provides a transparency and robustness that is not present in similar past frameworks.

The concept of the Planetary Boundaries has been widely taken up by the academic and policy community as the most robust planetary limits published thus far. Rockström et al.'s original (2009a) paper has over 3000 citations in 2019.

Not all reviews of the PBs are positive. The key criticisms are as follows:

- The limits proposed are based on a scientifically precautionary approach—a point of very low environmental risk. This approach may not seem fair to less developed nations who are balancing the need to manage environmental impacts with the need to provide citizens with basic needs (Galaz 2014; Galaz et al. 2012a).
- The Boundaries will change over time because of advances in scientific knowledge and interactions between boundaries (Steffen et al. 2015).
- There remains a need for value judgement in determining what constitutes "low risk" under the precautionary approach (Van Vuuren et al. 2016).
- Information will need to be gathered from different agencies to monitor the Boundaries. Some agencies may be reluctant to share this information (Galaz 2014).
- The Earth-system processes selected have been questioned (Lewis 2012; Mario 2009) as has the existence of global limits for some of the processes (Bass 2009; Molden 2009).
- There is no mechanism through which to address social wellbeing (Biello 2012).
- There is no international organization to coordinate the range of international, cross-sectorial, and multi-organizational initiatives (what we later will explain is a "poly-scalar approach") that would be needed for humanity to operate within the Planetary Boundaries (Galaz 2014).

Despite these concerns the PB concept has been widely adopted. Several authors have placed the PBs into the context of the broader definition of sustainability—considering social equity as well as environmental impact as part of sustainability (Raworth 2012; Steffen and Stafford Smith 2013). Raworth's (2012) concept of "Donut Economics" is similar to the Tolerable Windows approach. She has proposed minimum societal thresholds that should be met within the Planetary Boundaries—a "safe and just space for humanity".

There have been studies which aimed to improve the assessments of individual boundaries (e.g. Carpenter and Bennett 2011; Gerten et al. 2013; Mace et al. 2014), and proposals for alternative boundary processes (Running 2012). Barnosky et al. (2012) reviewed and confirmed the plausibility of global-scale tipping points, and de Vries et al. (2013) and Van Vuuren et al. (2016) have proposed alternative approaches to manage the complex interactions between the boundaries.

Although not specifically referred to in the Sustainable Development Goals (SDGs), the concept was included in many of the SDG proposals including SDSN (2013), Griggs (2013), and UNEP (2013). The UN High-Level Panel on Sustainability final report recommends that the PBs should be linked with policy (UN 2012). The European Environment Agency (EEA) have identified the PBs as an environmental priority and proposed a vision that we should aim to be living within them by 2050 (EEA 2011).

Several authors have identified the PBs as an opportunity to use science to inform policy (e.g. Brito 2012; European Commission 2012; Symons and Karlsson 2015) and to form targets for Earth-system governance (Galaz et al. 2012a). In a survey of eight European countries, seven reported that they found the PBs useful and important (Pisano and Berger 2013). One study suggested that since the PBs are a synthesis of decades of research from fields related to Earth-system science, they could be viewed as an operationalization of the biogeophysical component of sustainable development (Galaz et al. 2012a). Others have highlighted the opportunity for the PBs to reform environmental governance at multiple scales (Galaz et al. 2012c; Cole et al. 2014; Akenji et al. 2016; Häyhä et al. 2016). The PBs have even sparked interest with religious groups; the Dalai Lama held a meeting to discuss the connections between choices and environmental consequences including the PBs (Galaz et al. 2012b).

The Planetary Boundaries are summarized in Table 3.1. There are nine critical Earth-system processes, and one or more global and/or regional control variables and limits have been proposed for eight of these.

Table 3.1 Summary of the planetary boundaries (from Steffen et al. 2015, Table 1)

Earth-system process	Control variable	Planetary boundary	Current value
Climate change	Atmospheric concentration of carbon dioxide Change in radiative forcing	$\leq$350 ppm $\leq$1 W/m^2	396.5 ppm[a] 2.3 W/m^2
Biodiversity loss	Global extinction rate	$\leq$10 E/MSY	100–1000 E/MSY
Nitrogen and phosphorous cycle	Reactive nitrogen removed from the atmosphere Phosphorus flowing into oceans	$\leq$62 Tg N/year $\leq$11 Tg P/year	150 Tg N/year 22 Tg P/year
Stratospheric ozone depletion	Stratospheric concentration of ozone measured in Dobson units (DU)	$\leq$5% below preindustrial levels (290 DU)	~200 DU over Antarctica in Austral spring
Ocean acidification	Mean saturation state with respect to aragonite in the oceans	$\geq$80% of the preindustrial level	84% of the preindustrial level
Freshwater use	Freshwater consumption	$\leq$4000 km^3/year	~2600 km^3/year
Change in land use	Area of forested land as a percentage of original forest cover	$\geq$75%	62%
Novel entities	NA	NA	NA
Atmospheric aerosol loading	Aerosol optical depth	Regional limit of $\leq$0.25	0.3 AOD over South Asian region

Notes:
• ppm stands for parts (of carbon dioxide) per million (parts of atmosphere)
• Radiative forcing is the change in energy flux in the atmosphere measured in Watts per square meter of Earth's surface area (W/m^2)
• Extinction rate is measured in the number of extinct species per million species per year
• Saturation state with respect to aragonite is an indicator of ocean acidity
• Aerosol optical depth is a measure of the fraction of sunlight that is absorbed or reflected—a value of 0 indicates perfectly clear skies—a value of 1 indicates no sunlight penetration
[a]This was the value in 2015 when the PBs were updated. The current value is 405.51 ppm (NOAA 2018)

Some of the PBs, such as those for biosphere integrity and climate change, pertain to Earth-system processes which do not behave in a linear way and are likely to have tipping points. At certain threshold levels, these non-linear processes tend to undergo abrupt and sometimes irreversible change (Rockström et al. 2009b). Other PBs, such as those for nitrogen, are not associated with tipping points. These PBs have been included because they undermine Earth-system resilience or increase the risk of reaching thresholds for other processes (Rockström et al. 2009a). Together, the PBs have been dubbed the "safe operating space" for humanity (Rockström et al. 2009a).

In 2009 when the PBs were conceived, three of the nine PBs had already been exceeded: for climate change, biodiversity loss, and nitrogen release (Rockström et al. 2009a). Findings of the 2015 update of the PBs were that we had also surpassed the limit for change in land use (Steffen et al.). We are not currently operating in the safe operating space. This means that we are at risk of changing the state of the planet from one that is favourable to humanity to one that is hostile.

3.6   Conclusion

The state of the planet during the Holocene is the only state that we know is conducive to modern cities and agriculturally settled humans. There is evidence that we are in the transition to the next geological epoch, the Anthropocene, the long-term state of which is yet to be determined. Human activity is likely to be the major determinant of the state of the Anthropocene.

To minimize risk to humanity, we should aim for the state of the planet in the Anthropocene to resemble that of the Holocene. The Planetary Boundaries are environmental limits that are unlikely to shift back to the Holocene kind of levels without human intervention. It follows that humans should aim to operate within these Boundaries and to seek to regenerate the environment where the limits have been exceeded.

The next chapter goes on to explore theories on global environmental management to show how we might begin to operate within the Planetary Boundaries and to regenerate Earth systems for the future.

References

Akenji L, Bengtsson M, Bleischwitz R, Tukker A, Schandl H (2016) Ossified materialism: introduction to the special volume on absolute reductions in materials throughput and emissions. J Clean Prod 132:1

Alcott B (2010) Impact caps: why population, affluence and technology strategies should be abandoned. J Clean Prod 18:552–560

Azar C, Holmberg J, Lindgren K (1996) Socio-ecological indicators for sustainability. Ecol Econ 18:89–112

Barnosky AD, Hadly EA, Bascompte J, Berlow EL, Brown JH, Fortelius M, Getz WM, Harte J, Hastings A, Marquet PA, Martinez ND, Mooers A, Roopnarine P, Vermeij G, Williams JW, Gillespie R, Kitzes J, Marshall C, Matzke N, Mindell DP, Revilla E, Smith AB (2012) Approaching a state shift in Earth's biosphere. Nature 486:52–58

Bass S (2009) Planetary boundaries: keep off the grass. Nat Rep Clim Change 3:113

Berger A, Loutre MF (2002) An exceptionally long interglacial ahead? Science 297:1287–1288

Biello D (2012) Walking the line: how to identify safe limits for human impacts on the planet. Available: http://www.scientificamerican.com/article/do-planetary-boundaries-help-humanity-manage-environmental-impacts/. Accessed 8 Jun 2016

Bocquet-Appel J-P (2011) When the world's population took off: the springboard of the neolithic demographic transition. Science 333:560

Brito L (2012) Analyzing sustainable development goals. Science (New York, NY) 336:1396

Broman G, Holmberg J, Robört K-H (2000) Simplicity without reduction: thinking upstream towards the sustainable society. Interfaces 30:13–25

Carpenter SR, Bennett EM (2011) Reconsideration of the planetary boundary for phosphorus. Environ Res Lett 6:014009

Cohen JE (1995) How many people can the earth support? Norton, New York, NY

Cole MJ, Bailey RM, New MG (2014) Tracking sustainable development with a national barometer for South Africa using a downscaled "safe and just space" framework. Proc Natl Acad Sci U S A 111:E4399–E4408

Cook M (2005) A brief history of the human race. Granta, London

Crutzen PJ (2002) Geology of mankind. Nature 415:23

Daly HE (1990) Toward some operational principles of sustainable development. Ecol Econ 2:1–6

De Vries W, Kros J, Kroeze C, Seitzinger SP (2013) Assessing planetary and regional nitrogen boundaries related to food security and adverse environmental impacts. Curr Opin Environ Sustain 5:392–402

EEA (2011) Roadmap to a resource efficient Europe. European Environment Agency, Copenhagen

Ehrlich PR (1971) The population bomb. Pan Books, London

Ehrlich PR, Holdren JP (1971) Impact of population growth. Science 171:1212–1217

Ekins P, Simon S, Deutsch L, Folke C, De Groot R (2003) A framework for the practical application of the concepts of critical natural capital and strong sustainability. Ecol Econ 44:165–185

European Commission (2012) Proposal for a decision of the European parliament and of the council on a General Union Environment Action Programme to 2020 "Living well, within the limits of our planet". Off J Eur Union L354:171

Ewen C (2017) Oldest Homo sapiens fossil claim rewrites our species' history. Nature News

Ewing B, Moore D, Goldfinger S, Ourslet A, Reed A, Wackernagel M (2010) Ecological footprint atlas 2010. Global Footprint Network, Oakland, CA

F.N.L.P (1962) The collected letters of Antoni van Leeuzwenhoek. Edited, illustrated and annotated by a Commission of Dutch Scientists. Vol. VI. Amsterdam: Swets and Zeitlinger Ltd., 1961; pp. 12, 425; 34 plates. C8 15s. od. Med Hist 6:198

Fahrig L (2001) How much habitat is enough? Biol Conserv 100:65–74

Franck S, Von Bloh W, Müller C, Bondeau A, Sakschewski B (2011) Harvesting the sun: new estimations of the maximum population of planet Earth. Ecol Model 222:2019–2026

Galaz V (2014) Global environmental governance, technology and politics: the Anthropocene gap. Edward Elgar, Cheltenham

Galaz V, Biermann F, Crona B, Loorbach D, Folke C, Olsson P, Nilsson M, Allouche J, Persson Å, Reischl G (2012a) 'Planetary boundaries'—exploring the challenges for global environmental governance. Curr Opin Environ Sustain 4:80–87

Galaz V, Biermann F, Folke C, Nilsson M, Olsson P (2012b) Global environmental governance and planetary boundaries: an introduction. Ecol Econ 81:1

Galaz V, Crona B, Österblom H, Olsson P, Folke C (2012c) Polycentric systems and interacting planetary boundaries - emerging governance of climate change-ocean acidification-marine biodiversity. Ecol Econ 81:21–32

Galli A, Wackernagel M, Iha K, Lazarus E (2014) Ecological footprint: implications for biodiversity. Biol Conserv 173:121–132

Gerten D, Hoff H, Rockström J, Jägermeyr J, Kummu M, Pastor AV (2013) Towards a revised planetary boundary for consumptive freshwater use: role of environmental flow requirements. Curr Opin Environ Sustain 5:551–558

Global Footprint Network (2012) Glossary. Available: http://www.footprintnetwork.org/en/index. php/GFN/page/glossary

Global Footprint Network (2019) Footprint calculator. Available: http://www.footprintcalculator. org/. Accessed 24 Feb 2019

Goodland R, Daly H (1996) Environmental sustainability: universal and non-negotiable. Ecol Appl 6:1002–1017

Griggs D (2013) Sustainable development goals for people and planet. Nature 495:305–307

Häyhä T, Lucas PL, Van Vuuren DP, Cornell SE, Hoff H (2016) From planetary boundaries to national fair shares of the global safe operating space—how can the scales be bridged? Glob Environ Chang 40:60–72

Hoekstra AY (2009) Human appropriation of natural capital: a comparison of ecological footprint and water footprint analysis. Ecol Econ 68:1963–1974

IPCC (2007) Climate change 2007: the physical science basis. Contribution of working group I to the fourth assessment report of the intergovernmental panel on climate change. Cambridge University Press, Cambridge; New York, NY

IPCC (2013a) Climate change 2013: the physical science basis. Contribution of working group I to the fifth assessment report of the intergovernmental panel on climate change. Cambridge University Press, Cambridge; New York, NY

IPCC (2013b) Summary for policymakers. In: Stocker TF, Qin D, Plattner G-K, Tignor M, Allen SK, Boschung J, Nauels A, Xia Y, Bex V, Midgley PM (eds) Climate change 2013: the physical science basis. Contribution of working group I to the fifth assessment report of the intergovernmental panel on climate change. Cambridge University Press, Cambridge; New York, NY

Jouzel J, Masson-Delmotte V (2007) EPICA Dome C Ice Core 800KYr deuterium data and temperature estimates. Supplement to: Jouzel, Jean; Masson-Delmotte, Valerie; Cattani, Olivier; Dreyfus, Gabrielle; Falourd, Sonia; Hoffmann, G; Minster, B; Nouet, J; Barnola, Jean-Marc; Chappellaz, Jérôme A; Fischer, Hubertus; Gallet, J C; Johnsen, Sigfus J; Leuenberger, Markus; Loulergue, Laetitia; Luethi, D; Oerter, Hans; Parrenin, Frédéric; Raisbeck, Grant M; Raynaud, Dominique; Schilt, Adrian; Schwander, Jakob; Selmo, Enrico; Souchez, Roland; Spahni, Renato; Stauffer, Bernhard; Steffensen, Jørgen Peder; Stenni, Barbara; Stocker, Thomas F; Tison, Jean-Louis; Werner, Martin; Wolff, Eric W (2007). Orbital and millennial Antarctic climate variability over the past 800,000 years. Science, 317(5839), 793–797, doi: 10.1126/science.1141038. PANGAEA

Kukla GJ, Mattews RK, Mitchell JM Jr (1972) The end of the present interglacial. Quatern Res 2:261–269

Lewis S (2012) We must set planetary boundaries wisely. Nature 485:417

Lewis S, Maslin M (2015) Defining the Anthropocene. Nature 519:171–180

Mace GM, Reyers B, Alkemade R, Biggs R, Chapin FS, Cornell SE, Díaz S, Jennings S, Leadley P, Mumby PJ, Purvis A, Scholes RJ, Seddon AWR, Solan M, Steffen W, Woodward G (2014) Approaches to defining a planetary boundary for biodiversity. Glob Environ Chang 28:289–297

Malthus T (1798) An essay on the principle of population. J. Johnson, London

Margules CR, Nicholls AO, Pressey RL (1988) Selecting networks of reserves to maximise biological diversity. Biol Conserv 43:63–76

Mario JM (2009) Planetary boundaries: identifying abrupt change. Nat Rep Clim Change 1:115

Meadows DH, Meadows DL, Randers J, Behrens WW III (1972) The limits to growth: a report for the Club of Rome's project on the predicament of mankind. Universe Books, New York, NY

Molden D (2009) Planetary boundaries: the devil is in the detail. Nat Rep Clim Change 1:116

NOAA (2018) Trends in atmospheric carbon dioxide. Available: https://www.esrl.noaa.gov/gmd/ccgg/trends/monthly.html. Accessed 15 May 2018

Nordhaus WD, Tobin J (1971) Is growth obsolete? Cowles Foundation for Research in Economics, Yale University

Odum EP (1989) Ecology and our endangered life-support systems. Sinauer, Stanford, CT

Oxford Dictionaries (2014) English dictionary. Oxford University Press, Oxford. Available: http:// www.oxforddictionaries.com/. Accessed 16 Jul 2014

Paul JC (2002) Geology of mankind. Nature 415:23

Petschel-Held G, Schellnhuber HJ, Bruckner T, Tóth FL, Hasselmann K (1999) The tolerable windows approach: theoretical and methodological foundations. Clim Change 41:303–331

Pisano U, Berger G (2013) Planetary boundaries for SD: from an international perspective to national applications. European Sustainable Development Network, Vienna

Raworth K (2012) A safe and just space for humanity: can we live within the doughnut? Oxfam discussion paper. Oxfam, Oxford

Rees WE (1992) Ecological footprints and appropriated carrying capacity: what urban economics leaves out. Environ Urbaniz 4:121–130

Robèrt KH, Broman GI, Basile G (2013) Analyzing the concept of planetary boundaries from a strategic sustainability perspective: how does humanity avoid tipping the planet? Ecol Soc 18:5

Rockström J (2010) Let the environment guide our development. TED

Rockström J, Steffen W, Noone K, Persson A, Chapin FS, Lambin E, Lenton TM, Scheffer M, Folke C, Schellnhuber HJ, Nykvist B, De Wit CA, Hughes T, Van Der Leeuw S, Rodhe H, Sorlin S, Snyder PK, Costanza R, Svedin U, Falkenmark M, Karlberg L, Corell RW, Fabry VJ, Hansen J, Walker B, Liverman D, Richardson K, Crutzen P, Foley J (2009a) Planetary boundaries: exploring the safe operating space for humanity. Ecol Soc 14:32

Rockström J, Steffen W, Noone K, Persson Å, Chapin FS, Lambin EF, Lenton TM, Scheffer M, Folke C, Schellnhuber HJ, Nykvist B, De Wit CA, Hughes T, Van Der Leeuw S, Rodhe H, Sörlin S, Snyder PK, Costanza R, Svedin U, Falkenmark M, Karlberg L, Corell RW, Fabry VJ, Hansen J, Walker B, Liverman D, Richardson K, Crutzen P, Foley JA (2009b) A safe operating space for humanity. Nature 461:472–475

Running S (2012) A measurable planetary boundary for the biosphere. Science 337:1458–1459

SDSN (2013) An action agenda for sustainable development. Report for the UN Secretary-General. Sustainable Development Solutions Network, Paris

Severinghaus JP, Sowers T, Brook EJ, Alley RB, Bender ML (1998) Timing of abrupt climate change at the end of the Younger Dryas interval from thermally fractionated gases in polar ice. Nature 391:141

Soulé ME, Sanjayan MA (1998) Ecology: conservation targets: do they help? Science (New York, NY) 279:2060

Spreckley F (1987) Social audit: a management tool for co-operative working. Beechwood College, Leeds

Steffen W, Stafford Smith M (2013) Planetary boundaries, equity and global sustainability: why wealthy countries could benefit from more equity. Curr Opin Environ Sustain 5:403–408

Steffen W, Sanderson A, Tyson P, Jäger J, Matson P, Moore B, Oldfield F, Katherine R, John Schellnhuber H, Turner BL, Wasson RJ (2005) Global change and the earth system: a planet under pressure. Springer, Berlin

Steffen W, Richardson K, Rockström J, Cornell SE, Fetzer I, Bennett EM, Biggs R, Carpenter SR, De Vries W, De Wit CA, Folke C, Gerten D, Heinke J, Mace GM, Persson LM, Ramanathan V, Reyers B, Sörlin S (2015) Planetary boundaries: guiding human development on a changing planet. Science 347:1259855

Steffen W, Rockström J, Richardson K, Lenton TM, Folke C, Liverman D, Summerhayes CP, Barnosky AD, Cornell SE, Crucifix M, Donges JF, Fetzer I, Lade SJ, Scheffer M, Winkelmann R, Schellnhuber HJ (2018) Trajectories of the Earth system in the anthropocene. Proc Natl Acad Sci 115:8252–8259

Symons J, Karlsson R (2015) Green political theory in a climate-changed world: between innovation and restraint. Environ Polit 24:173–192

The Brundtland Commission (1987) Our common future. World Commission on Environment and Development. Oxford University Press, Oxford

Trenberth K (2002) Volume 1, The Earth system: physical and chemical dimensions of global environmental change. In: MacCracken MC, Perry JS (eds) Encyclopedia of global environmental change. John Wiley & Sons, Ltd, Chichester

UN (2012) Resilientpeople, resilient planet: a future worth choosing. United Nations, New York, NY

UNEP (2013) Embedding the environment in sustainable development goals. In: UNEP post-2015 discussion paper 1. United Nations Environment Programme, Nairobi

Vale R, Vale B (2013) Living within a fair share ecological footprint. Routledge, New York, NY

Van Vuuren DP, Lucas PL, Häyhä T, Cornell SE, Stafford-Smith M (2016) Horses for courses: analytical tools to explore planetary boundaries. Earth Syst Dynam 7:267–279

Vitousek PM, Mooney HA, Lubchenco J, Melillo JM (1997) Human domination of Earth's ecosystems. Science 277:494–499

Wackernagel M, Rees WE (1996) Our ecological footprint: reducing human impact on the earth; illustrated by Phil Testemale. New Society, Philadelphia, PA

Wackernagel M, Schulz NB, Deumling D, Linares AC, Jenkins M, Kapos V, Monfreda C, Loh J, Myers N, Norgaard R, Randers J (2002) Tracking the ecological overshoot of the human economy. Proc Natl Acad Sci U S A 99:9266–9271

WBGU (1995) Scenario for the derivation of global CO2-reduction targets and implementation strategies. WBGU, Bremerhaven

WBGU (2001) Conservation and sustainable use of the biosphere. Earthscan, London

WBGU (2005) World in transition: fighting poverty through environmental policy. Earthscan, London

WBGU (2006) The future oceans – warming up, rising high, turning sour. WBGU, Berlin

WBGU (2011) World in transition: a social contract for sustainability. WBGU, Berlin

WBGU (2014) Human progress within planetary guardrails: a contribution to the SDG debate. Policy paper 8. WBGU, Berlin

Zalasiewicz J, Williams M, Haywood A, Ellis M (2011) The Anthropocene: a new epoch of geological time? Philos Trans R Soc A Math Phys Eng Sci 369:835–841

Chapter 4
Managing the Earth System: Why We Need a Poly-Scalar Approach

Ruin is the destination toward which all men rush, each
pursuing his own best interest in a society that believes in the
freedom of the commons
Hardin (1968, p. 1244)

Abstract Human activity is altering critical natural processes beyond global limits. The way we manage the global environment over the next few decades will be a major determinant for the state of the Earth for the next epoch—the "Anthropocene". Current efforts at managing the environment often use a top-down approach, an idea based on out-of-date theories of environmental management. Efforts at different levels are often piecemeal, with no cohesion or common direction beyond a general goal of reducing environmental impacts. Moreover, most people have a relatively small sphere of influence. Global environmental problems can feel overwhelming— it can be difficult to see how the efforts at smaller scales could influence global outcomes.

Three areas of social science, observed human behaviour, commons management, and change theory, can be used to show that a different approach is needed. These theories highlight the benefits of breaking down our global environmental problems into smaller pieces so that they can be tackled at the scale of magnitude that is most effective. They show the value of an approach that is integrative across different scales of society (i.e. individual, community, city, small and large business, and government at all scales) and across different timescales (from short term to long term). Further, they demonstrate the importance of an approach that is not overly prescriptive—that defines the "what" but allows people to determine the "how". Finally, they convey the necessity that the approach should instil trust in others that people are working to the same end.

We thus propose a new "poly-scalar" approach that uses a general system of rules to generate trust that others are working towards the same goal. The system of rules needs to have the resolution to communicate what is needed at different scales of magnitude, the flexibility to let people determine how to achieve this, and the mechanisms to enable the integration of different timescales and scales of activity

© Springer Nature Singapore Pte Ltd. 2020
K. Meyer, P. Newman, *Planetary Accounting*,
https://doi.org/10.1007/978-981-15-1443-2_4

in society. It is our position that a poly-scalar approach would enable more effective management of the global environment. It could help us to return to and live within the planet's limits as it links Earth sciences with the social sciences.

4.1 Introduction

Despite global consensus that we are changing the state of the planet, we are failing to mitigate our behaviour at a rate befitting the urgency of the problem. As outlined in Chap. 3, we have already exceeded four Planetary Boundaries—safe environmental limits for humanity (Steffen et al. 2015). There is an urgent need to manage global-scale impacts on the environment from human activity and even create urban and agricultural systems that begin regenerating the planet (Newman et al. 2017; Thomson and Newman 2016, 2018).

The problem is how. The phenomenon of our poor ability to manage shared resources known as "the commons" is not new. The idea can be traced back as far as 350 BC to Aristotle (1996, Politics, Book II, Chapter 3, p. 33). Lloyd predicted irresponsible use of shared property as early as 1833 (Lloyd 1977). These early struggles to manage shared resources were typically at the scale of a single forest or fishery. Now humanity is faced with the challenge of managing the global environment—a magnitude of environmental problems that is overwhelming to most people.

The purpose of this chapter is to introduce a new theory of how the Earth system could be managed. Commons management theories suggest that the most effective approach to managing the environment would be one which can be applied at different scales of magnitude and using different decision-making frameworks. Observed human behaviour change shows that pro-environmental decisions rely on an overlap of community-, government-, and business-driven parameters. Change theory highlights the importance of overlapping interests across different timescales and emphasizes the need to integrate different areas of society—community, government, and businesses.

This chapter introduces these theories and shows how they can be brought together to make a case for a poly-scalar approach to Earth-system management.

4.2 Theories of Commons Management

Theories about how to manage the commons began to appear in academic literature in the 1950s and 1960s (Olson 1965; Hardin 1968; Gordon 1954). These early theories described in the following sections resulted in an assumption that either privatization or top-down governance was needed to manage the commons. There is still a prevalence in top-down governance despite advances in commons management theories (see Box 4.1).

Box 4.1 Is the Earth System Really a "Commons"?

Should commons theory be used as the basis for Earth-system management? Is the Earth system really a commons? A *common pool resource* or *common goods* are defined as resources or goods of which:

1. The exclusion of some users is difficult.
2. The use of the resource or good is subtractible—the use of the good or resource by one person reduces the capacity for others to use it.

For example, Hardin's open pasture with limited grazing is a commons. As the pasture is open, it therefore meets the exclusion criteria. The more cows that graze the pasture, the less grass per cow. The resource is thus subtractible.

The Earth system is not by definition a commons. It meets the first criteria. However, it does not always meet the second. A person breathing from the atmosphere does not reduce the capacity for others to do the same. Officially the Earth system is defined as a "public good". Environmental impacts, for example, pollution, are referred to as "public bads".

It is generally agreed by experts in the field that the behavioural dynamics around public goods, public bads, and commons are similar and that common pool resource theory can thus be applied to all three (Ostrom et al. 1961; Lo and Tang 1994; Gardner et al. 2000).

4.2.1 Conventional Theories on Managing the Commons

The most famous of the conventional theories is Hardin's "tragedy of the commons" (1968). Hardin explained his theory of the commons using the example of an open pasture for cattle. Before the pasture reaches capacity, each cow grazed on the pasture can be sold for the value of 1. Once the pasture reaches capacity, an additional cow added to the pasture would result in a loss to the value of $1/x$ per cow, where x is the total number of cows. A farmer adding a cow beyond the pasture's capacity would thus receive almost the full gain from the additional cow, while the losses from overgrazing would be distributed evenly across all the cows and therefore shared by every farmer. Hardin argued that the rational decision for each farmer is thus to add more cows, until such point that the pasture can no longer support any cows at all. Herein lies the tragedy (1968).

A variant of the tragedy of the commons was proposed by Olson, "Collective Action" (1965). Olson postulates:

Unless there is coercion or some other special device to make individuals act in their common interest, rational self-interested individuals will not act to achieve their common or group interests. (Olson 1965, p. 2)

His theory was not specifically about commons, but rather group behaviour in general, and has led to further work on related topics such as voluntary compliance and lobbying group activities (Rupasingha and Boadu 1998).

Both theories are based more broadly on the "Prisoner's dilemma". Two (guilty) prisoners are simultaneously questioned. There are four possible outcomes:

1. Both prisoners remain silent; both face 1 year in prison.
2. Both prisoners testify; both face 2 years in prison.
3. Prisoner A testifies against Prisoner B, while Prisoner B remains silent; Prisoner A is set free, Prisoner B faces 3 years in prison.
4. Prisoner B testifies against Prisoner A, while Prisoner A remains silent; Prisoner B is set free, Prisoner A faces 3 years in prison.

The game does not allow for any communication between prisoners and assumes there is no risk of future consequences of ousting the other prisoner. Thus, according to the rules of the game, both prisoners will testify, even though this leads to the worst possible outcome (i.e. a total of four prisoner years).

One may argue that group logic should prevail that the farmers should realize the tragedy and coordinate their behaviour. The counterargument to this is that in any group attempting to act for the greater good, there is a risk of "free riders". Free riders are actors who reap the long-term benefits of positive action by others without contributing to the upfront costs (Ostrom 1990). The existence or even anticipation of free riders has been found to influence others' behaviour towards individual gain rather than group wellbeing (Olson 1965; Ostrom 1990).

It is easy to see how these theories could be applied to our response to so many of our environmental dilemmas. For example, we know that greenhouse gas levels are well above safe limits and that to avoid severe temperature increases, we must stop emitting them. Yet, if you consider an individual deciding whether to drive or take the train on a rainy day, one can understand how she might feel that the potential environmental benefits of this one train ride would have little impact given the scale of the problem. Moreover, if she looked out the window and saw many "free riders" staying dry in the comfort of their cars, she would likely feel even less motivated to make the effort for this small payoff.

Those who subscribe to the theory of the commons typically propose either the privatization or the top-down governance of shared resources (Ostrom 1990). There are some very successful examples of both management structures. The Montreal Protocol is considered by most to be a highly successful example of global top-down governance (Epstein et al. 2014), although it could also be argued that it is not an example of top-down governance at all (see Sect. 4.5.1).

However, there are constraints and barriers in any management structure. In the example of the atmosphere—there have been several decades of attempted global, top-down governance to manage climate change. Yet, the development of a global solution has been problematic with many aspects that are hotly debated including the magnitude of emission reductions required, the methods and strategies to achieve emission reductions, and the division of responsibilities and costs. The recent Paris Agreement can be viewed as a substantial achievement in global policy (Schleussner et al. 2016). However, the nationally determined contributions under the agreement are estimated to correlate to warming of 2.7–3 °C, well over the Paris target of 1.5 °C (Schleussner et al. 2016; Sharma 2017).

4.2.2 Modern Theories on Managing the Commons

In the 1980s and 1990s, there was a substantial body of theoretical literature that contradicted the arguments of the conventional theories, including by Hardin himself (Hardin 1982; Marwell and Ames 1980; Ostrom et al. 1994; Runge 1981, 1984; Sandler 1992; White and Runge 1994).

Nobel Prize winner Elinor Ostrom began a movement in 1990 which disputed the validity of the theory of the commons altogether. She showed through empirical evidence that the theory that individuals and small groups will not change their behaviour without external enforceable rules is far from inevitable. There are many studies showing examples of well-managed shared resources such as forests and fisheries (Ostrom 1990, 1988; Bernard and Young 1997; Freeman 1989; Korten 1987; Korten and Klauss 1984; McCay and Acheson 1987; National Research Council 1986; Siy 1982). In some instances these self-organized regimes have proved more effective than would have been feasible in the case of privatization or top-down governance (McKean 1998; Ostrom 2010) (see Box 4.2).

Box 4.2 Alanya Fishery

Ostrom uses a fishery in Alanya, Turkey, to demonstrate how self-governance can be more effective than privatization or top-down governance.

The fishery of approximately 100 fishers was in a bad way from overharvesting. Conflict among fishers was high.

The local co-operative spent a decade of trial and error and came up with a system whereby each year the top fishing spots were identified and agreed upon, and a roster was then made so that each fisher rotated through the fishing spots. The system was governed by the fishers themselves. The fishery thrived. The conflict subsided.

A government official or private organization could not have derived such a solution. The solution relied on the in-depth knowledge of the area of the local co-operative. It was made economically viable by the ability of the fishers, who were already on location, to self-enforce the rotation.

Studies of observed human behaviour show that the overarching factors which lead to co-operative behaviour by individuals towards the commons are:

- The development of trust that the behaviour will lead to long-term benefits even if there are short-term costs
- The belief that the majority of actors are performing this behaviour (Ostrom 2009)

"Keeping up with the neighbours", a study rolled out by the Sacramento Municipal Utility District, is a good example of how trust in others' actions can promote positive behaviour change. Personalized reports of power consumption with comparisons to their neighbours' consumption were given out to a portion of the residents. Those who received the personalized reports reduced power consumption by significantly more than those who did not (Kaufman 2009).

There are several key aspects which can make the self-regulation of commons preferable to the privatization or governance of these resources:

1. Users of a resource tend to have a breadth and depth of knowledge of its constraints and opportunities that would be extremely time and cost intensive to obtain externally. This knowledge may allow for the development of more effective resource management schemes than would be possible otherwise (Ostrom 1990).
2. The time and cost implications of the enforcement of policies have often led to failed regulation by government and private organizations. Self-organizing groups have the unique opportunity for self-regulation, with all actors playing the part of both user and regulator, thus reducing the complexity and cost of policy enforcement (Ostrom 1990).
3. In many instances the removal of ownership from users of the commons takes away the sense of stewardship and leads to increased exploitation of the resource (McKean 1998; Ostrom 1990).

Of course, there are many instances where self-regulation of resources has failed. It is not enough to rely on self-regulation to manage the Earth system. Management theory suggests that there is not a single solution to managing the environment. It is a complex system and requires a flexible and adaptable approach that is likely to include governance, privatization, and self-regulation.

When considering the management of the Earth system, it is important to consider not only the type of management or decision-making framework (e.g. self-organized, privatized, governed, etc.) but also the scale of management (e.g. local, national, global, etc.). Formal attempts at managing the Earth system, particularly with respect to climate change, have thus far been predominantly at a global level, although there are notable exceptions such as C40 Cities which is an example not only of formal action at a smaller scale but also of self-organized management of the global commons (C40 2017). While some form of global scale agreement will almost certainly be a necessary component of successful management of the Earth system, this is not the only scale that is important.

Global environmental problems are typically caused by a multitude of actions which take place at a far smaller scale (Ostrom 2009; Kates and Wilbanks 2003). Given the diverse nature of the causes of climate change, global or even national policies could miss many opportunities for emission reductions. Further, trust is often greater in local bodies than in national governments as local bodies are perceived to have more awareness of local conditions.

Small scale or local initiatives alone would of course be insufficient to manage a problem such as climate change as many opportunities to reduce emissions rely on decisions which can only be made at a larger scale (Kates and Wilbanks 2003). One might also argue that there are already many small-scale actions taking place with limited global success.

4.2.2.1 Polycentric Management

Research into the governance of metropolitan areas has shown that while large-scale governance systems are an essential element in the efficient management of cities and metropolitan areas, small- and medium-scale components were also necessary. Ostrom draws parallels from this to climate change mitigation (2009). She suggests that there is no one solution for the management of the commons and that each case should be considered individually. In a background paper to the *World Development Report* for the World Bank, Ostrom (2009) proposed a polycentric approach for dealing with climate change, where a polycentric order is defined as:

> one where many elements are capable of making mutual adjustments for ordering their relationships with one another within a general system of rules where each element acts with independence of other elements. (Ostrom 1999)

Polycentric systems are characterized by multiple governing authorities at different scales, in contrast to a monocentric unit (Ostrom 2010). The general system of rules is included as a mechanism to impart trust in the long-term benefits of the actions and that others are contributing to the same goal (see Box 4.3).

Box 4.3 A General System of Rules to Impart Trust in Others
In most places, households are free to operate as they see fit. There is not an external party suggesting what the rules should be in your household, defining when you should sleep or what you should eat. Yes, there is a general system of rules—laws, social norms, cultures—that you and your neighbours most likely adhere to. You dispose of your waste according to the general rules— you do not throw your rubbish into your neighbour's yard. You can feel confident that they will do the same. There is a sense of trust that others will generally abide by the general system of rules.

There are many studies which support Ostrom's proposal for a polycentric approach (e.g. Neuvonen et al. 2014; Galaz et al. 2012; Brondizio et al. 2009). Reports on global biodiversity management refer to the importance of action at all levels and across different decision-making centres (Secretariat of the CBD 2014, 2006, 2010). Agenda 21, the outcome of the 1992 Earth Summit, identifies public participation as a "fundamental prerequisite" for sustainable development. It proposes national government collaboration with local government and "major groups" (e.g. women and indigenous people) (United Nations Conference on Environment and Development (UNCED) 1992). Local Agenda 21 initiatives emerged throughout the world, some of which are ongoing today (Wittmayer et al. 2016).

Variations of a polycentric approach have proven successful in other applications. Lorek and Spangenberg (2001) identified a need for actor-centric measurement as a means to change behaviour. This is consistent with Holmberg et al.'s (1999) finding that quantifying individual environmental impacts helps to motivate people to reduce their consumption. This phenomenon has been shown to be effective, for example, in driving improvement in healthcare (Meyer and Merry 2017) (see Box 4.4).

Box 4.4 A Polycentric Approach to NZ Healthcare

In response to the large numbers of falls reported annually to the NZ Health Quality and Safety Commission (the Commission), a programme was developed to reduce harm from falls. Rather than the traditional top-down approach, the Commission's approach was multilayered. Support and guidance (or "the general system of rules") comes from the centre. However, the implementation was flexible, focussed on every individual context, and driven from those on the ground. Every clinician was expected to ask at every encounter with a patient, "How can this person be prevented from falling—and what can I personally do to achieve this?" The Commission found that providing feedback to the clinicians that allowed them to track their own performance and compare it to their counterparts was a very effective means of driving change. Recently, high levels of compliance with the local process indicators of the falls programme have been matched with a statistically significant reduction in the rate of hip fractures (the national-level outcome indicator), which is a substantial achievement in respect of a clinical challenge known internationally to be difficult. While NZ healthcare may seem very distant to the management of our global environment, the learning outcomes of this example are highly relevant. Specifically, giving individuals the opportunity to determine the best method of achieving a specified outcome was shown to be very effective in driving change.

Aggregated marginal gains is the idea that if every possible trivial gain is made, these multiple little successes add up to provide worthwhile benefit. This has been used with high success to raise the British Cycling team from underdogs to Tour de France winners in only 3 years (Myers 2016). The concept of aggregated gains has been advanced into theories of agglomerated economies. Agglomeration is about more than multiplying activities; it is also about the scale at which certain functions in society work best. As Ostrom points out, for management to be effective, it must occur at a scale that "can encompass the problem" (Ostrom et al. 1961). This concept has been applied to designing sustainable cities where agglomeration economies are understood to be the basis of wealth creation and productivity gains due to the meaningful overlap of skills and integration of networks that are critical to the knowledge economy (Glaeser and Gottlieb 2009; Graham and Dender 2011; Trubka 2011).

A global environment management system should thus have the resolution to coordinate different magnitudes and types of activity in a way that instils trust that others are working towards the same end goal. Such a system would enable us to leverage off local knowledge while at the same time forming the basis of global negotiation. It would help to mitigate the influence of free riders and allow us to benefit from the efficiencies borne by approaching environmental problems at the scale at which they can best be addressed.

4.3   Theories of Behaviour Change

Managing human impacts on the environment means managing human behaviour—whether this is individual day-to-day behaviour or behaviour as a CEO, as an innovator, or as a government official.

Early theories of behaviour change were based on the idea that people behave rationally and that behaviour was predictable. For example, the theory of planned behaviour is that one's beliefs are linked to one's behaviour—that if someone has the intention to perform a behaviour that they will do so. The theory of social norms is that a person's behaviour is influenced by their perception of others' behaviour—that they will be more likely to perform a behaviour if they believe that this is what is expected of them. The theory of cognitive dissonance is that people aim to be consistent with their attitudes, beliefs, and actions.

These early theories led to the belief that the most effective ways to change behaviour were through the provision of information and feedback about the behaviour, as well as by influencing social norms.

More recent studies, based on observed human behaviour, have shown that behaviour is very difficult to predict. Decisions vary with lifestyle, position within a family or within society, attitudes, motivations, habits, knowledge, past behaviours, social norms, context, and technology (Eon et al. 2017, 2018). Traditional approaches—such as identifying positive past behaviours to help a person to strengthen their own identity as someone who cares for the environment, can help

(Steg 2016); however motivation and identity on their own are insufficient in driving change. Practice theory, for example, highlights the importance of addressing *context* and *technology* as well as motivations, knowledge, and habits (Eon et al. 2017). In a framework for encouraging pro-environment behaviour, Steg and Vlek (2009) identify the three key factors which influence environmental behaviour as motivation, context, and habit.

The critical takeaway from these latest findings is the role of *context* and *technology* in decision-making. This means that to begin to change behaviour, one must look more broadly than individual and community values and norms, motivations, habits, and knowledge. The roles of business (technology) and government infrastructure and regulations (context) must also be considered. For example, smart phone apps which give live updates regarding bus and train timetables have been shown to significantly increase the use of public transport, i.e. a technology has enabled a behaviour change (Newman and Kenworthy 2015). Similarly, the provision of segregated bins for recycling increases recycling rates.

Thus, according to behaviour change theory, an effective approach to achieving pro-environmental behaviour is likely to be one which encompasses community values and norms, business innovations and technology, and government regulations and infrastructure.

Fig. 4.1 The magic of sustainability occurs when community, business, and government interests overlap. (Adapted from Newman and Rowe 2003)

4.4 Change Theory and Sustainability

Newman (2005) writes about the "magic" of sustainability. He describes the magic as an innovative solution which exceeds expectations. He muses that the magic most often appears a result of a new, integrative approach to problem-solving that allows space for reflexive learning. Specifically, he suggests that when community values and visions overlap with government regulations and infrastructure, and business innovations—whether products or services—this is when the magic of sustainability can occur (see Fig. 4.1).

An important element of this concept is that it brings together different timescales. The values and visions of a community are typically tied into culture and beliefs—which stretch over a long timescale. Parents have dreams for the future for their children, grandchildren, and even great grandchildren. Government timescales are typically mid-term. Infrastructure projects such as rail systems and buildings are built to last over decades. Government strategies are often prepared with a 20- to 30-year view. On the other hand, governments often have high inertia and are slow to make change. Projects can be under consideration for decades prior to their implementation. In contrast, businesses operate rapidly, moving from one innovation to another. They can be agile and flexible. There is space for trial and error. Businesses have a very short-term focus. Stakeholders demand rapid payback. It is not uncommon for business decisions to be made with consideration only of the implications over the next few years. Corporate sustainability plans frequently have targets spanning less than a decade.

When these groups come together, businesses can begin to innovate with the long-term visions and aspirational goals of the community in mind. Governments can gain from the fast-pace approach of businesses while incorporating the long-term views of the community. When there is space to discover an overlap of common interests, the magic can occur.

This notion of using integration to obtain a result that is better than the sum of its parts is not new, nor is it limited in its success to sustainability applications. For example, "holistic design" is an architectural concept that has become popular over the last decade. Holistic design is synonymous with integrated design and occurs when all stakeholders are included in the design process from the outset. Innovative solutions borne out of such an approach often achieve efficiencies that would be unlikely without the involvement of all parties from the start.

An integrative approach to finding solutions is just one component of change theory. In his book *The Tipping Point*, Gladwell (2000) evaluates why some innovations generate change, while others with seemingly similar potential fail. He suggests that there are several common characteristics associated with successful innovations: *the law of few, the stickiness factor, and context*. The importance of context to drive behaviour change has already been discussed in Sect. 4.3. The stickiness factor relates to the message or product's memorability, the ability to "stick" in one's mind. The third factor, the law of few, is that social epidemics are led by a few key people. These people are commonly referred to in the literature as agents of change.

Agents of change, sometimes referred to as "local heroes" in sustainability circles, can be individuals operating in a community, CEOs driving change through business, or government officials who create change through their position in society. For example:

> Rosa Parks was an individual who by refusing to give up her seat in the white area of the bus inspired community action that was a key component in the black rights movement in the United States.
> Winston Churchill took over the leadership of Great Britain during the Second World War and encouraged the country to stand up to Hitler, a move that was instrumental to the Allied war effort victory in 1945.
> Elon Musk has been instrumental in accelerating the global transition to sustainable energy.

The power of the individual is lost in a top-down governance model that suggests that to persuade people to act for the greater good, we must impose regulations. Global societal changes such as the end of slavery, black rights, and women's rights have come about through collective bottom-up efforts at varying scales towards a singular goal.

In summary, change theory highlights the importance of integration across sectors and timeframes. It again highlights the critical role of innovation, technology, and context. Finally, it confirms the importance of different scales of action, in particular the scale of the individual, in driving change.

4.5 Poly-Scalar Management of the Earth System

The three theories described above can be used to advance Ostrom's proposal for a polycentric approach. Polycentricity at its core is about multiple scales of governance (Ostrom et al. 1961; Ostrom 2010). Ostrom has identified the importance of different decision-making frameworks which include self-organized initiatives, privatization, and governance (Ostrom 1990, 1988; Ostrom et al. 1994). Yet, her discussions about the benefits of a polycentric approach refer predominantly to the inclusion of local and regional scales of governance (Ostrom 2010, 2009, 1999). Further, a polycentric approach does not demand integration across different sectors, nor does it consider different timeframes.

As shown in this chapter, theories of behaviour, of commons management, and of change point to the importance of an integrated approach—not only of different scales of governance from local to global but also of different decision-making frameworks including self-regulation, privatization, and governance, across different sectors of society from community, to business, and government and of different timescales from the short term to long term.

We therefore propose a new approach to managing the Earth system: a poly-scalar approach. We define this as a management system which is:

integrative across different scales of magnitude, timescales, and scales of society, that is coordinated by a general system of rules which have different mechanisms at different centres of activity.

There are many benefits of such a high-resolution approach.

Immediacy: it is unlikely that a global agreement can be made quickly regarding a top-down approach to global environmental management. There is not time to wait for global agreement. A poly-scalar approach does not rely on a global approach. Efforts at different scales can thus begin straight away. This does not preclude the continued efforts to agree to top-down solutions—which will almost certainly play a critical element in successful management of the Earth system. Further, a poly-scalar approach would not mean that all people would immediately change. However, it would enable those ready to act to do so in a more cohesive way without delay.

Cost-effectiveness: The development of a global solution would require a high level of certainty of the efficacy of the solution. This would mean costly analysis and assessment of various options. In contrast, using a poly-scalar approach would provide the opportunity for trial and error, results of which could be fed back into the development of larger-scale actions and policies (Ostrom 1990; Kates and Wilbanks 2003). Capturing lessons learnt from a wide range of approaches would facilitate a high rate of knowledge uptake at a relatively high speed and low cost.

Linking scales: Solutions to our environmental problems are likely to require changes at every level, from the day-to-day activities of individuals, families, and communities to the policies and regulations of companies and nations. A poly-scalar approach could be used to connect these scales of activity to global outcomes. The benefits of mitigating climate change will also occur at different levels. For example, a household choosing to invest in insulation and energy efficient appliances will see long-term payback through reduced energy bills; the inhabitants of a city in which the use of cars is minimized will all reap the benefits of the cleaner air.

Integrative thinking: Bringing together local heroes, community values, national legislation, and business drivers may help to develop solutions that are better than the sum of their parts. Local communities and cities are in a position to identify specific opportunities for improvements that might not be obvious at a larger scale. However, it is easier to hold cities or nations (as opposed to individuals) accountable if they take on the role of a free rider. In the transport example, a nation may have targets to reduce transport-related impacts in order to meet international commitments. However it is most likely a city-level decision to provide a comprehensive public transport solution and therefore give individuals the opportunity to make a lower impact transport solution (Newman and Kenworthy 2015). At a local level, people would reap the benefits of less congestion, reduced noise pollution, and improved air quality. The decisions of the individuals, city planners, and national leaders would, together, contribute to the national and global targets.

Building trust: A high-resolution approach that could be applied transparently across different scales of activity would help to build trust at all scales that others are working to the same end, thus mitigating the effects of free riders.

Table 4.1 shows examples of different areas, scales, and sectors that would be encompassed in a poly-scalar approach.

4.5.1   The Montreal Protocol: A Successful Example of Top-Down Global Governance or of a Poly-Scalar Approach?

As previously stated, the Montreal Protocol is often used as an example of successful top-down governance. However, it could be argued that its success in fact lies in its poly-scalar approach.

It was first recognized that some substances could and were depleting the ozone layer in the mid-1970s (Chesick 1975). In 1985 an article was published in *Nature* confirming that there was a repeating springtime hole in the ozone layer (Farman et al. 1985). The Montreal Protocol was first ratified in 1989 (UNEP 2017a). In 2009 it became the first treaty to have universal ratification (UNEP 2017b). Scientific evidence shows that the ozone hole is reducing (Solomon et al. 2016). Models predict that provided the Montreal Protocol continues to be followed, the hole will continue to recover (Solomon et al. 2016).

The hole in the ozone layer is a global-scale environmental problem that is the result of a public bad (the pollution of the atmosphere with ozone depleting substances). On the face of it, it appears that the public bad was resolved through top-down governance—a global treaty for change.

However, there are some important factors believed to have contributed to the protocol's success:

Table 4.1 Examples of different areas of activity in a poly-scalar approach to Earth-system management

Scale	Areas of activity		
Large	Global community groups: e.g. *IPCC, WWF*	Global governance: e.g. *United Nations*	Multi-national firms: e.g. *Unilever, Mars*
Medium	City—national-scale community groups, e.g. *ACF, YCA, C40 Cities*	National, state, city government	Medium-sized businesses
Small	Households, communities, neighbourhoods	Local government	Small businesses
Individual	Individuals, local heroes	PMs, mayors, local heroes	CEOs, sustainability managers, employees
Sector (timeframe)	**Community (long-term visions and values)**	**Government (medium-term focus)**	**Business (agile, short-term outlook)**

- Prior to the development of the protocol, there were growing community groups lobbying heavily for the removal of the substances that were causing rapid sunburn among children in southern latitudes (Stocker et al. 2012).
- The Montreal Protocol included mechanisms to target not only policy-makers but also institutions who gained the most from these substances (Parson 2003).
- The key manufacturers of the chemicals involved in the Montreal Protocol already had a preferable solution to the problem and therefore would not get in the way of regulations to ban their chemicals. It simply was a problem of governments making mechanisms to phase out their use in existing products like spray cans and refrigerants.

These factors show evidence of integration across different scales and sectors, the important role technology played, and the presence of change agents pushing community values and visions. Figure 4.2 shows how these factors fit into the "magic of sustainability" concept from Fig. 4.1.

The Montreal Protocol is an example of a poly-scalar approach to managing an Earth-system process.

Fig. 4.2 The magic of the Montreal Protocol

4.6   Conclusion

Past theories led to the belief that privatization and top-down governance were the only effective ways to manage shared resources. However, the latest social science knowledge, from the fields of behaviour change, commons management, and change theory, suggests that an alternative approach would be preferable.

In this chapter we have demonstrated a need for a poly-scalar approach to Earth-system management. We have shown how a general system of rules can be used to generate trust that others are working towards the same end. We have also shown that a poly-scalar approach can be used to break down global problems into a scale of magnitude that is manageable and to develop solutions that integrate different timescales and scales of society.

In the next chapter, we explore accounting theory and how this can be used to guide the design of the general system of rules that would be needed if we were to use such an approach that would enable us to return to and live within the Planetary Boundaries.

References

Aristotle (ed) (1996) The politics and the constitution of Athens. Cambridge University Press, Cambridge

Bernard T, Young J (1997) The ecology of hope: communities collaborate for sustainability. New Society Publishers, Gabriola Island, BC

Brondizio ES, Ostrom E, Young OR (2009) Connectivity and the governance of multilevel social-ecological systems: the role of social capital. Annu Rev Env Resour 34:253

C40 (2017) C40 cities. Available: http://www.c40.org/. Accessed 28 Jul 2017

Chesick JP (1975) Atmospheric halocarbons and stratospheric ozone. Nature 254:275

Eon C, Morrison GM, Byrne J (2017) Unraveling everyday heating practices in residential homes. Energy Procedia 121:198–205

Eon C, Morrison G, Byrne J (2018) The influence of design and everyday practices on individual heating and cooling behaviour in residential homes. Energ Effic 11:273–293

Epstein G, Pérez I, Schoon M, Meek CL (2014) Governing the invisible commons: ozone regulation and the Montreal Protocol. Int J Commons 8:337–360

Farman JC, Gardiner BG, Shanklin JD (1985) Large losses of total ozone in Antarctica reveal seasonal ClOx/NOx interaction. Nature 315:207–210

Freeman DM (1989) Local level organizations for local development: concepts and cases of irrigation organization. Westview Press, Boulder, CO

Galaz V, Crona B, Österblom H, Olsson P, Folke C (2012) Polycentric systems and interacting planetary boundaries - emerging governance of climate change-ocean acidification-marine biodiversity. Ecol Econ 81:21–32

Gardner R, Herr A, Ostrom E, Walker JA (2000) The power and limitations of proportional cutbacks in common-pool resources. J Dev Econ 62:515–533

Gladwell M (2000) The tipping point. Little Brown, New York, NY

Glaeser EL, Gottlieb JD (2009) The wealth of cities: agglomeration economies and spatial equilibrium in the United States. J Econ Lit 47:983–1028

Gordon HS (1954) The economic theory of a common-property resource: the fishery. J Pol Econ 62:124–142

Graham D, Dender K (2011) Estimating the agglomeration benefits of transport investments: some tests for stability. Plan Pol Res Pract 38:409–426

Hardin G (1968) The tragedy of the commons. Science 162:1243–1248

Hardin R (1982) Collective action. Johns Hopkins University Press, Baltimore, MD

Holmberg J, Lundqvist U, Robèrt KH, Wackernagel M (1999) The ecological footprint from a systems perspective of sustainability. Int J Sustain Dev World Ecol 6:17–33

Kates R, Wilbanks T (2003) Making the global local responding to climate change concerns from the ground. Environ Sci Pol Sustain Dev 45:12–23

Kaufman L (2009) Utilities turn their customers green, with envy. The New York Times. 31 January, 2009

Korten D (1987) Introduction: community-based resource management. In: Korten D (ed) Community management: Asian experience and perspectives. Kumarian Press, Hartford, CT

Korten D, Klauss R (1984) People centred development: contributions toward theory and planning frameworks. Kumarian Press, Hartford CT

Lloyd W (1977) On the checks to population. In: Hardin G, Baden J (eds) Managing the commons. Freeman, San Francisco, CA

Lo CW, Tang SY (1994) Institutional contexts of environmental management: water pollution control in Guangzhou, China. Public Admin Dev 14:53–64

Lorek S, Spangenberg J (2001) Environmentally sustainable household consumption. Wuppertal Institute, Wuppertal

Marwell G, Ames RE (1980) Experiments on the provision of public goods. II. Provision points, stakes, experience, and the free-rider problem. Am J Sociol 85:926–937

McCay BJ, Acheson JM (1987) The question of the commons. The culture and ecology of communal resources. University of Arizona Press, Tucson AZ

McKean M (1998) Common property: what is it, what is it good for, and what makes it work? In: Gibson C, McKean M, Ostrom E (eds) Forest resources and institutions. The Food and Agriculture Organization of the United Nations, Rome

Meyer K, Merry A (2017) Saving civilization through personal budgeting in a quality improvement paradigm. In: Herzberg A (ed) Statistics, science and public policy, vol XXI. National Library of Canada Cataloguing in Publication, Hailsham

Myers J (2016) This coach improved everything by 1%, putting Britain on the road to Rio Olympic glory. World Economic Forum. Accessed 9 Sep 2019

National Research Council (1986) Proceedings of the Conference on Common Property Resource Management. National Academy Press, Washington, DC

Neuvonen A, Kaskinen T, Leppänen J, Lähteenoja S, Mokka R, Ritola M (2014) Low-carbon futures and sustainable lifestyles: a backcasting scenario approach. Futures 58:66–76

Newman P (2005) Can the magic of sustainability revive environmental professionalism? Greener Manag Int 49:11–23

Newman P, Kenworthy J (2015) The end of automobile dependence: how cities are moving beyond car-based planning. Island Press, Washington, DC

Newman P, Rowe M (2003) The Western Australian state sustainability strategy: a vision for quality of life in Western Australia. Department of the Premier and Cabinet, Perth, WA

Newman P, Beatley T, Boyer H (2017) Resilient cities: overcoming fossil fuel dependence. Island Press, Center for Resource Economics, Washington, DC

Olson M (1965) The logic of collective action: public goods and the theory of groups. Harvard University Press, Cambridge, MA

Ostrom E (1988) The rudiments of a theory of the origins, survival and performance of common property institutions. In: Korten D (ed) Making the commons work. Kumarian Press, Hartford DC

Ostrom E (1990) Governing the commons: the evolution of institutions for collective action. Cambridge University Press, Cambridge; New York, NY

Ostrom V (1999) Polycentricity—Part 1. Polycentricity and local public economies. University of Michigan Press, Ann Arbor, MI

Ostrom E (2009) A polycentric approach for coping with climate change. World Bank, Washington, DC

Ostrom E (2010) Polycentric systems for coping with collective action and global environmental change. Glob Environ Chang 20:550–557

Ostrom V, Tiebout CM, Warren R (1961) The organization of government in metropolitan areas: a theoretical inquiry. Am Pol Sci Rev 55:831–842

Ostrom E, Gardner R, Walker J (1994) Rules, games, and common-pool resources. University of Michigan Press, Ann Arbor, MI

Parson EA (2003) Protecting the ozone layer: science and strategy. Oxford University Press, New York, NY

Runge CF (1981) Common property externalities: isolation, assurance, and resource depletion in a traditional grazing context. Am J Agric Econ 63:595–606

Runge CF (1984) Institutions and the free rider: the assurance problem in collective action. J Polit 46:154–181

Rupasingha A, Boadu FO (1998) Evolutionary theories and the community management of local commons: a survey. Rev Agric Econ Agric Appl Econ Assoc 20:530–546

Sandler T (1992) Collective action: theory and applications. University of Michigan Press, Ann Arbor MI

Schleussner CF, Rogelj J, Schaeffer M, Lissner T, Licker R, Fischer EM, Knutti R, Levermann A, Frieler K, Hare W (2016) Science and policy characteristics of the Paris agreement temperature goal. Nat Clim Change 6:827–835

Secretariat of the CBD (2006) Global biodiversity outlook 2. Convention on Biological Diversity, Montreal, QC

Secretariat of the CBD (2010) Global biodiversity outlook 3. Convention on Biological Diversity, Montreal, QC

Secretariat of the CBD (2014) Global biodiversity outlook 4. Convention on Biological Diversity, Montreal, QC

Sharma A (2017) Precaution and post-caution in the Paris agreement: adaptation, loss and damage and finance. Clim Pol 17:33–47

Siy RY (1982) Community resource management: lessons from the Zanjera. University of the Philippines Press, Manila

Solomon S, Ivy DJ, Kinnison D, Mills MJ, Neely RR, Schmidt A (2016) Emergence of healing in the Antarctic ozone layer. Science 353:269–274

Steffen W, Richardson K, Rockström J, Cornell SE, Fetzer I, Bennett EM, Biggs R, Carpenter SR, De Vries W, De Wit CA, Folke C, Gerten D, Heinke J, Mace GM, Persson LM, Ramanathan V, Reyers B, Sörlin S (2015) Planetary boundaries: guiding human development on a changing planet. Science 347:1259855

Steg L (2016) Values, norms, and intrinsic motivation to act proenvironmentally. Annu Rev Env Resour 41:277–292

Steg L, Vlek C (2009) Encouraging pro-environmental behaviour: an integrative review and research agenda. J Environ Psychol 29:309–317

Stocker L, Newman P, Duggie J (2012) Climate change and Perth (South West Australia). In: Blakely E, Carbonell A (eds) Resilient coastal regions: planning for climate change in the United States and Australia. Lincoln Institute of Land Policy, Washington, DC

Thomson G, Newman P (2016) Geoengineering in the Anthropocene through Regenerative Urbanism. Geosciences 6:46

Thomson G, Newman P (2018) Urban fabrics and urban metabolism: from sustainable to regenerative cities. Resour Conserv Recycl 132:218–229

Trubka RL (2011) Agglomeration economies in Australian cities: productivity benefits of increasing urban density and accessibility. Doctor of Philosophy, Curtin

UNEP (2017a) The Montreal protocol on substances that deplete the ozone layer. Available: http://ozone.unep.org/en/treaties-and-decisions/montreal-protocol-substances-deplete-ozone-layer. Accessed 22 Dec 2017

UNEP (2017b) Treaties and decisions. Available: http://ozone.unep.org/en/treaties-and-decisions. Accessed 22 Dec 2017

United Nations Conference on Environment and Development (UNCED) (1992) Agenda 21. UNCED, New York, NY

White TA, Runge CF (1994) Common property and collective action: lessons from cooperative watershed management in Haiti. Econ Dev Cult Change 43:1–41

Wittmayer JM, Van Steenbergen F, Rok A, Roorda C (2016) Governing sustainability: a dialogue between local Agenda 21 and transition management. Local Environ 21:939–955

Chapter 5
Environmental Accounting, Absolute Limits, and Systemic Change

If I had asked people what they wanted, they would have asked for a faster horse
Henry Ford

Abstract The level of change needed for humanity to operate within the planet's limits will almost certainly require fundamentally altering the way we interact with the environment. We need urgent, systemic change.

Accounting theory shows that standards and limits are needed to create serious change. Measuring the changing state of the environment (environmental assets) and human pressures on the environment (environmental flows) is known as environmental accounting. Environmental accounting is used by many countries, cities, regions, and businesses as a tool to help them understand and manage their environmental impacts. However, there is a major limitation in most environmental accounts—the lack of appropriate limits.

Environmental performance targets are typically set based on incremental targets. These are often set based on feasibility, local policies, or industry best practice. They are arbitrary. Achieving industry best practice standards does not equate to maintaining or improving environmental assets. Incremental targets encourage incremental change.

In contrast, absolute limits that defined the end goal for our environmental assets and flows would help us to understand what is needed. Absolute limits can be used to create a vision of the future that is not constrained by the status quo.

The Planetary Boundaries are absolute environmental limits. What is missing is a mechanism to connect these limits to existing environmental accounting practices. A connection is suggested as part of Planetary Accounting. This connection could be used to create a design brief for the future—a platform for disruptive innovation and transformational change.

© Springer Nature Singapore Pte Ltd. 2020
K. Meyer, P. Newman, *Planetary Accounting*,
https://doi.org/10.1007/978-981-15-1443-2_5

5.1 Introduction

Accounting theory highlights the importance of measuring and monitoring assets and flows in order to make informed decisions. Governments, private organizations, and households alike make informed decisions and choices based on their knowledge of the state of their assets and of incoming and outgoing cashflow. Environmental accounting translates these insights from accounting theory to the management of environmental impacts.

Environmental impact assessment (EIA)—the quantification of environmental damage from human activity—was first formalized in 1969 at the United Nations Conference on the Environment in Sweden (Biswas and Modak 1999). It was first introduced into government legislation the following year by the US National Environmental Policy. By the early 1990s, EIA was part of national legislation for more than 20 nations (Biswas and Modak 1999).

The translation of EIA into environmental accounting—the practice of measuring and monitoring environmental assets, gains, and losses over time—followed quickly. Norway was one of the first countries to begin keeping formal environmental accounts. They identified their environmental assets—forests, fisheries, energy, and land—and began to track the state of these in the early 1980s (Saebo 1994). The Netherlands introduced the National Accounting Matrix which included environmental accounts in 1991 (De Boo et al. 1993; Biswas and Modak 1999; BIS 2012).

In response to the Rio Earth Summit in 1992, the United Nations developed the System of Environmental-Economic Accounting (SEEA) (UN 1993). It was developed in collaboration with the World Bank, to assess the feasibility of using monetary accounting practices to assess natural resources. The most recent update—the SEEA Central Framework—was adopted by the UN Statistical Commission as the first international standard for environmental-economic accounting (UN Statistics Division 2018).

In the 1990s, new methods for assessing environmental impacts on a global and local scale were developed such as the Ecological Footprint (Rees 1992). Some of these are used to estimate the impacts of different scenarios to inform decision-making (e.g. Global Footprint Network 2014), but as is showed below, this technique is not very scientific and has lost a lot of credibility to use in national accounting. Today, environmental accounting refers to the practice of assessing past, current, and estimated future environmental impacts and reporting the results over time, often against specified targets. We assess *environmental assets* (the state of the environment) and *environmental flows* (human appropriation of environmental goods and services). These techniques continue to evolve and are the basis of this work for Planetary Accounting though much more focussed.

This chapter begins by introducing two different types of change, systemic and incremental. This is followed by an overview of EIA methods and the benefits and limitations of environmental accounting. The concept of absolute limits is then introduced. Accounting theory and carbon accounting are used to show why absolute limits are essential if we are to generate systemic change and be successful in managing human impacts on the environment.

5.2 Systemic vs. Incremental Change

In Chap. 3 we argued that humans need to return to and live within the Planetary Boundaries (PBs). Four of the nine Planetary Boundaries have already been exceeded. These are not marginal transgressions. The current rate of species extinction is between 10 and 100 times the PB for biosphere integrity. The energy imbalance in the atmosphere is more than double the "safe" level for climate change. More than twice the PB limits for phosphorus and almost three times the PB limits for nitrogen are being released into the environment each year. Almost one billion hectares of forest needs to be replanted—an area approximately the same size as the United States—to meet the PB for land use.

To meet the needs of the projected (and current) global population without reducing Earth's capacity to support the way of life that many of us in the richest nations have now come to expect is not a small undertaking. It may not be possible at all. It is unlikely that it will be possible without fundamentally rethinking the way humans operate—i.e. without systemic change.

Incremental change is typically the preferred path for decision-makers. Incremental change refers to small changes and adjustments—usually to elements of a larger system. For example, to improve the energy efficiency of a light bulb the bulb needs to be changed; however the rest of the system (e.g. the wiring and switches) can remain untouched. Incremental change does not threaten existing models and frameworks. The implications of incremental change can usually be predicted with a reasonable level of confidence. In contrast, systemic change means change to the entire system (Oxford Dictionary 2018). It is often referred to as transformative or transformational change (Termeer et al. 2017; Seijts and Gandz 2018). Systemic change is not simply a large increment of change. Further, incremental change does not usually lead to systemic change though as Gladwell (2000) suggests tipping points can be small and unpredictable.

Henry Ford, who was the first to mass produce automobiles, purportedly said, "If I had asked people what they wanted, they would have asked for a faster horse". Too often, efforts towards managing the environment focus on reducing impacts—the faster horse or incremental approach. Consider modern cars, for example.

In light of the increased understanding that people must emit less carbon dioxide to the atmosphere, many companies have worked hard to improve the fuel economy of their petrol and diesel vehicles—i.e. to reduce the carbon dioxide emissions per kilometre of travel: incremental change. The greater the efficiency becomes, the harder it becomes for the companies to find ways to further improve it. The invention of the electric car is game-changing. It puts the whole system of petrol vehicles, petrol stations, and fossil fuel transport in question. Suddenly companies who have invested so much time and energy into improving the efficiency of their petrol engines are competing with zero emission vehicles.

Electric cars were invented before the petrol engine (Bellis 2017) (see Fig. 5.1). However, the first modern electric car that could compete with modern petrol vehicles—i.e. to drive at comparable speeds and over 100+ kilometres in a single charge—the Tesla Roadster, was not released to the public until 2008 (see Fig. 5.1). The Nissan Leaf, the first mass market electric vehicle produced by a major manufacturer, was released in 2010. Only 8 years later, China, India,

Fig. 5.1 1915 Detroit electrical vehicle (top) and tesla roadster 2.5 (Sfoskett 2005 and Overlaet 2011, CC BY SA 3.0, CC BY-SA 3.0: Creative Commons License allows reuse with appropriate credit)

France, Britain, and Norway are working towards phasing out petrol and diesel cars altogether (Gray 2017). The system is changing. Car companies who do not quickly make the transition to electric vehicles risk going out of business[1]. In contrast, companies which had the foresight to develop electric vehicles have also had a head start in thinking about related business opportunities such as charging stations and in-home charging equipment. The invention of electric cars is an example of disruptive change (see Chap. 5 that is likely to lead to systemic change. However, the story does not end here.

There are experts who predict that the role of the automobile needs to and will diminish, e.g. Newman and Kenworthy (2006) and Newman and Kenworthy (2015). These authors have identified that cities built around cars are less amenable and less economic to their inhabitants than those which are not. They have shown that increasingly, cities are shifting towards alternative modes of transport. The authors propose alternative, more efficient uses for land area that is currently used for roads such as knowledge economy, economic activity, and public open space to help with walkability. Ending automobile dependence is an example of true systemic change. This is an example of the extent of change that may be required to end the global environmental crisis.

The fundamental problem with incremental targets is that they can often be met with incremental changes, at least to begin with. This means that a lot of effort can go into improving the efficiency of fundamentally inefficient systems. Innovative solutions can be missed with such an approach (Akenji et al. 2016). For example, if car companies considered the magnitude and breadth of change required to transition into a sustainable future—they might conclude that the demand for any form of private vehicle is on the decline, or is likely to be soon, and shift their focus from low emission or electric vehicles to the development of innovative semiprivate or public transportation solutions.

There is a phenomenon known as the rebound effect, or moral licencing, which can occur as a result of incremental reductions to environmental impacts. For example, using a low energy car more than before, or following the installation of photovoltaic (solar) panels on a house, a surge in electricity consumption is often found. This is because the occupants feel that they have made good steps towards reducing impacts on the environment and can therefore relax about sustainable habits such as turning off unneeded lights. This phenomenon is typically association with incremental improvements rather than systemic change (Arvidsson et al. 2016; Hertwich 2005; Kojima and Aoki-Suzuki 2015).

To live within the Planetary Boundaries, we will need to create systemic change—the problem is: How to achieve this?

---

[1] Elon Musk has made the patents from the Tesla vehicle available to anyone. Musk, E. 2014. *All Our Patent Are Belong To You* [Online]. Tesla. Available: https://www.tesla.com/en_NZ/blog/all-our-patent-are-belong-you [Accessed 22 April 2018]. Electric vehicle companies who have been slow to transition may thus still be able to keep up with the transition to electrical vehicles. This is an unusual occurrence, but a good example of how an individual can act as a change agent to achieve global change (see Chap. 5).

5.3   Environmental Impact Assessments: A History

In the early days of environmental impact assessment, this constituted measuring the state of the environment. For example, the total area of forest and the number of trees per meter squared might have been monitored.

Over time, people began to consider not only the state of the environmental assets, but also the environmental flows, which effected the state of the assets (like cashflow in economic accounting). In a forest, this could have meant monitoring and recording the number of trees planted, the number of trees extracted, the amount of wood produced or wasted, the amount of fertilizer being applied, and so forth. The SEEA Central Framework now sets out how to monitor environmental assets and flows so that these are monitored in a consistent way between different nations (UN 2014).

In the early days of environmental accounting, the environmental limits were often clear. In the example of a forest, people understood rates of growth for different tree species and could thus determine the maximum rates of extraction that could occur without diminishing the forest. However, as we began to monitor more complex systems, the limits became less obvious. For example, we know now that releasing sulphate into the atmosphere contributes to air pollution. Yet, if only a small amount of sulphate was released, the change to the atmospheric conditions would be negligible. No one has determined a safe limit for local or global sulphate emissions. This is true for many environmental flows with a few notable exceptions such as the release of carbon dioxide into the atmosphere.

Despite a lack of upper limits, environmental accounting is common practice for many businesses, cities, and nations and can also be done for individuals, groups of people, or products and services. There are many types of environmental impact assessments, e.g. life-cycle assessment (LCA), environmental footprint assessment, and material flow accounting (MFA). LCA and environmental footprint assessment are two of the most commonly used types and are therefore described in more detail in the following sections.

5.3.1   Life-Cycle Assessments

LCA is the process of identifying and tabulating all resource and waste streams from the extraction of raw materials, the production of the product, the use over its lifetime, the transportation of the raw material and the product, and finally the disposal.

The earliest studies, which were not then called life-cycle assessments, but are now considered to have been partial life-cycle assessments, date back to the late 1960s and early 1970s (Guinée 2012). These assessments were typically undertaken to compare different packaging options. Initially, only energy consumption was considered. Later, resource use, waste, and emissions were also taken into account (Guinée 2012).

Until the 1990s, LCAs were performed using different methods, terminologies and approaches. In 1994 the International Organization for Standardization (ISO) produced the first international standard for life-cycle assessments: ISO:14040:1997 Environmental management—Life-cycle assessment—Principles and framework (ISO 1997). This standard enabled a much greater level of consistency, robustness, and transparency in life-cycle-assessments.

Many environmental claims made about products pertain to only a single aspect of a product's environmental impacts. For example, cars are often compared to one another for their fuel economy. Yet a car with better fuel economy may or may not have lower impacts over its lifetime. The overall impacts depend on a multitude of factors, for example, the amount of raw materials used, the efficiency of the manufacturing processes, or the overall lifespan of the car. LCA allows the big picture of environmental impacts to be better understood. LCAs take every aspect of a product's environmental impact into account, thus allowing a far greater resolution for comparing the different impacts of a product or service. Moreover, LCA includes a mechanism with which to weight and aggregate the different impacts into a single score so that overall impacts of one product can be easily compared to another.

There are three major limitations of LCA:

1. To do a formal LCA assessment requires vast amounts of data that can be costly in both resources and time to acquire (Kirchain Jr. et al. 2017).
2. Aggregating different environmental impacts into a single score is imprecise and can give meaningless results (Kalbar et al. 2017).
3. Without aggregating results into a single score, the results of LCA can be difficult to communicate to the layperson (Kalbar et al. 2017).

The LCA community are active in looking at both product- and organization-level impacts. However, the limitations described above have led to a relatively poor uptake of LCA by businesses despite the depth of detail an LCA can provide (Kalbar et al. 2017).

5.3.2 Environmental Footprints

At approximately the time that LCA methods were being formalized into ISO standards, the concept of the Ecological Footprint was published (Rees 1992). An Ecological Footprint (EF) is a measure of the natural capital used for or by a system (e.g. a person, group, or product). The results are expressed in a proprietary unit—global hectares (gha). Global hectares are a weighted unit of area to allow different land types to be compared for equivalent environmental value. For example, 1 ha of forest land was considered equivalent to 1.2 gha in 2006 (Valada 2010), i.e. 1 ha of forest was estimated to be 20% more productive than a world average hectare (Galli et al. 2007; Monfreda et al. 2004). The EF of a product, person, or jurisdiction can be compared to the corresponding biological capacity (biocapacity). Biocapacity is a measure of available natural capital, also expressed in gha.

The EF does not account for all human impacts; however the authors attempted to capture all impacts which compete for space in this metric (Galli et al. 2014). The amount of each land type (see Table 5.1) used for a given activity is tabulated to determine the total EF of the activity. CO_2 emissions are included in the EF through an equivalent forest area—the area of forest that would be needed to absorb that amount of CO_2. This is not directly scientific as each form of energy has clearly got impacts that are not simply transferrable to an area of land, e.g. wind power would be the same as coal depending just on how much it is consumed.

The EF was not intended as a stand-alone sustainability indicator (Ewing et al. 2010a). It does not account for:

1. Availability or depletion of non-renewable resources.
2. Inherently unsustainable activities (e.g. wastes which the biosphere has no assimilative capacity for, such as the release of heavy metals, radioactive compounds and persistent synthetic compounds).
3. Environmental management and harvest practices.
4. Land and ecosystem degradation (yield factors do not take into account sustainability of practices).
5. Ecosystem disturbance or resilience of ecosystems.
6. Use or contamination of freshwater.
7. Non-CO_2 greenhouse gases.

Further, the authors have identified key limitations of the framework as:

- Aggregation of different land types giving the impression of more equivalence between different land types than is realistic.
- Accuracy is limited by the quality of datasets, many of which are incomplete and do not include confidence limits.
- The methodology leads to underestimation of the extent of impacts.

Notwithstanding the above, the concept of the Ecological Footprint quickly became very popular (see Chap. 3). There are online calculators with which individuals can estimate their own footprint (Global Footprint Network 2014). There is a formal methodology for calculating national Ecological Footprint accounts (Ewing et al. 2010b) which has been used to calculate the Ecological Footprint for most countries (Global Footprint Network 2011).

Table 5.1 Biocapacity and Ecological Footprint land categories

Biological capacity	Ecological Footprint
• Cropland	• Cropland Footprint
• Grazing land	• Grazing Footprint
• Forest land	• Forest Product Footprint
• Fishing ground	• Fish Footprint
• Built up land	• Carbon Footprint
	• Built up land

The EF spawned many other footprint concepts including the water, carbon, and nitrogen footprints. Each footprint measures a specific impact of a person, group, activity, or product. The results can be expressed in a variety of units such as mass, volume, or area. In a review of footprint analysis tools, Cucek et al. (2012) identified 31 different environmental footprints.

Environmental footprints typically assess a single environmental impact such as the amount of carbon emissions or land used for a certain activity or set of activities. Environmental footprints can also include upstream and downstream impacts of an activity. However, the calculation process varies between footprints, and few footprint indicators are regulated by International Organization for Standardization (ISO) standards. This is limiting for their use with major firms seeking to do serious environmental accounting.

Footprints have been criticized for assessing only a single element of the environmental impacts of an activity. Some authors argue that footprints should not be considered as wholistic measures of environmental sustainability (Laurent and Owsianiak 2017). However others argue that the benefit of footprint assessments is that the results are easily communicated to the general public which is an important aspect of understanding environmental impacts (Ewing et al. 2010a). The issue of single environmental impact indicators is being addressed by several scholars through the use of footprint families (e.g. Fang 2015b; Fang 2015a; Fang et al. 2014; Galli et al. 2012). It would appear that the world is looking for something better to base their scientifically derived sense of environmental activity limits.

5.3.3   The Benefits of Environmental Accounting

Without environmental accounting, it can be very difficult to determine whether one activity, product, or behaviour is better or worse for the environment than another. Almost every trip to the supermarket, a consumer finds they are faced with the same problem—should they buy the tin of local-grown tomatoes or the organic tomatoes that are grown and tinned overseas and then imported. How do you know which has less impact on the environment? It would take a substantial amount of work to determine the answer. Even if the tomato companies reported their water, carbon, and nitrogen footprints or made a full set of LCA data available for the two products, the answer may still be unclear.

To add even more complexity, the way we purchase things also changes the overall impacts. If faced with the option to drive across town to a farmer's market where the consumer could buy local, organic, and fresh tomatoes instead of tinned tomatoes, the impacts of driving across town may render this a false economy. A study was done in Sweden where the greenhouse gas emissions associated with different groceries were estimated. What was interesting in this study is that the difference in impacts from local and seasonal food compared to the average bag of groceries was very similar to the difference in impacts between travelling 20 km per week for groceries compared to 40 km. The single greatest impact reductions were achieved through the decision to limit the purchase of beef (Baumann 2013).

We can thus conclude that measuring and understanding the environmental impacts of different decisions is necessary to enable people to make informed decisions about our interactions with the natural world. However, this information needs to be provided in a manner which is robust, transparent, and easily comprehensible.

5.3.4 Production Versus Consumption Accounting

As with any form of numerical modelling, the usefulness of the model depends greatly on the way the model is used. One key distinction between environmental accounts is whether they are considering the impacts related to consumption by a population or related to the production within a defined area (Wiedmann 2009). Both consumption and production accounts provide important and useful information. However, a common quirk in interpreting the results of environmental accounts is to use production accounts to derive consumption data. For example, national production accounts are frequently reported in per capita figures and compared to per capita data for other nations. Given the global distribution of energy, food, and other products and services, this sort of comparison provides very little meaningful data. The impact of a net-exporting country, such as Australia or New Zealand, will have higher production impacts per capita than a net-importing country such as Denmark or Singapore. This is not to say that Australians and New Zealanders do not consume more than the Danish or Singaporeans. It is quite possible that they do. The problem is that production accounts do not help us to answer this question.

5.4 Absolute Sustainability

Environmental accounting allows humans to take responsibility for managing our impacts on the environment. It is now possible to estimate ahead of time what the environmental impacts of different decisions might be. While the accuracy of such estimations is limited, these estimations can be useful to inform decision-making, planning, policy, and legislation.

As discussed previously, the shortcoming of environmental accounting is that results of environmental assessments are typically reported against self-selected targets. Targets are almost always based on improvements to the status quo or a past environmental state, rather than a desired future state founded in science (Akenji et al. 2016). Existing environmental assessment tools have been identified as lacking in suitability to inform society regarding environmental matters because of this lack of science-based targets or limits (Laurent and Owsianiak 2017; Akenji et al. 2016).

The Kyoto and Paris Agreement targets are an example of this. Most of the targets are set based on percentage reductions from a past benchmark—typically 2005 or 1990 levels of emissions. It is arbitrary how far below 1990 or 2005 emissions levels a countries' emissions might already be, and for the future it is just important to get started on *reducing* not increasing emissions. However we will need to eventually develop something more scientific if limits are to be taken seriously. Scientific methods can be used to predict with a reasonable level of confidence how many emissions of greenhouse gases we can afford globally while remaining below a given temperature target. It would be possible from this prediction to set a global, science-based target for emissions and then negotiate shares of this global budget. Of course, this would not resolve the difficulties of negotiating shares. However, with a clear target laid out, so too is the basis for systemic change. This is the Planetary Accounting preferred approach.

Incremental targets can lead to missing opportunities for systemic change (Akenji et al. 2016; Sandin et al. 2015). Incremental improvements are the basis of most personal and policy change (Newman et al. 2017) and the importance of these should not be overlooked. Systemic change can sometimes be achieved through incremental action. However, without understanding the end goal, incremental improvements are unlikely to lead to systemic change. To live within the Planetary Boundaries, reducing GHG emissions and designating protected zones to safeguard habitats might be necessary but will not be sufficient. Scientific assessments of necessary change and potential impacts and adaptations will be needed for all the planetary boundaries.

The idea of "absolute sustainability" or "absolute limits" refers to the idea of limits which are at a point that there are no longer negative impacts on the environment. Quantifying the point at which this occurs and being able to compare this to current impacts is considered by many to be critical in the management of human impacts (Akenji et al. 2016). It is believed by many experts that absolute limits will be needed to drive the systemic change necessary to transition to a sustainable future (Bahadur and Tanner 2014; Pelling et al. 2015). Others recommend that a sustainable future should be defined by absolute environmental limits rather than efficiency improvements from the status quo (Akenji et al. 2016; Fang et al. 2014).

In financial accounting, people do not make decisions based only on the state of one's assets and rate of outgoing cashflow compared to a benchmark or industry standard. This information is informative, but grossly lacking as the basis of financial management. An understanding of the maximum available cashflow is fundamental to managing accounts. To make informed environmental choices, governments, business, and individuals need to understand the maximum environmental capital available to them.

Many efforts targeted at increasing individual's motivations to reduce their impacts on the environment focus on scare tactics. Images of polar bears on ice-caps, predictions of devastating weather events, and warnings of impending doom are rife in the media. They have little impact. Most people feel stressed and overwhelmed by the news (Newman 2005). Some react by taking the less emotionally challenging viewpoint of sceptics (see Chap. 3). A more effective method is to provide a hopeful outlook: "If we do x, we can achieve y", and this can be phased and shared equally (Newman 2005). Absolute limits would enable the creation of visions of a future where we were no longer degrading our environment. They could be used to connect with people and create behaviour change. They would become the basis of political change, and the detail can be worked out by everyone at whatever scale they work at.

Absolute limits can also be used to create a design brief for future technology and innovation. Disruptive innovation is a term used to describe solutions that are driven dramatically by demand and displace the status quo of whatever system it is applied to (see Box 5.1). Disruptive innovations often (though not always) leave past systems obsolete. They are an important driver of change and almost always lead to systemic change. Smart phones are an example of disruptive innovation that like all disruptive change is not easily predicted as they grow so fast due to their demand rather than simple cost comparisons; Newman et al. (2017) suggests this is also now observable with solar PV systems and that systemic changes to grids are happening quite quickly in response. Uber is another example of disruptive innovation with systemic change; the most successful global taxi business does not own its fleet of vehicles though whether the whole transport system is disrupted can be debated. The photography industry is a further example. Photography businesses used to make most of their income developing pictures—they made a profit per photo taken. Then Kodak invented the digital camera and suddenly photos became free. This invention forced the photography industry to undergo systemic change— from a per photo income model to a tools and services business. Kodak changed the product but did not alter their systems quickly enough—they went out of business. Absolute limits would provide a platform for disruptive innovation as innovators would see what is needed to change and could dream of the potential to disrupt the system that is preventing change.

Box 5.1 Disruptive Innovations
Disruptive innovators understand that demand is not based purely on cost. People want things for many reasons. Some disruptive innovations take a top-down route—i.e. they are immediately superior solutions which are initially unaffordable, but the cost comes down over time because of demand. The Tesla vehicle is an example of an innovation that is likely to be a top-down, disruptive innovation. Some disruptive innovations are considered to be bottom-up—the solutions are initially inferior and unaffordable but, over time, become both superior and affordable. Photovoltaic (solar) panels are an example of a bottom-up innovation. The best, most disruptive innovations which are superior and cheaper from the start are known as big bang disruptions. Google maps is an example that made previous navigations obsolete almost overnight. And similarly smart phones became universal in a very short period and replaced the need for landline systems.

5.5   Carbon Accounting: Environmental Accounting with Absolute Limits

Carbon accounting (or GHG accounting) is the most widely used form of environmental accounting. Carbon accounting refers to the practice of measuring and reporting emissions of CO_2 and sometimes other GHGs. The important difference between carbon accounting and other environmental accounting practices is that the results can be easily compared to global limits.

There are debates as to a "safe" level of global warming and therefore maximum allowable CO_2 emissions. Nonetheless, it is possible to translate a global target of average global warming in degrees Celsius, to a corresponding concentration of CO_2 in the atmosphere, and then to a maximum budget for anthropogenic CO_2 emissions. CO_2 emissions for an activity can thus be linked to a global budget based on scientific knowledge a "science-based target". It is becoming increasingly common for cities, countries, and businesses to use science-based targets for greenhouse gases. For example, in New Zealand, a consortium of businesses have created the Climate Leader's Coalition—to become a member, businesses are required to set and publically announce science-based greenhouse gas targets. Carbon accounting has led to widespread understanding of what is a relatively complicated scientific problem.

Individuals can calculate their "carbon footprint"—the amount of CO_2 released due to the activities of the individual. Formal GHG accounting protocols have been developed for nations, cities, and products and services (Fong et al. 2014; Greenhalgh et al. 2005). CO_2 emissions have been translated into dollar values. Studies have been completed to assess the relative benefits of a carbon tax versus carbon trading. Different approaches for managing emissions and different technologies for reducing emissions or absorbing carbon from the atmosphere have been trialled in different locations and at different scales allowing for a very rapid uptake of knowledge and development. New forms of governance and behaviour as well as products, urban forms, and systems of professional activity have all been developed because the scientific information has been provided in clear and accountable formats.

Carbon accounting is a remarkable example of the importance of limits. Different scales and types of emissions management are taking place across the globe. These efforts at every scale have already led to some success. Economic growth has been decoupling from greenhouse gas emissions since 2000. For the third year in a row, population and GDP have increased, while global CO_2 emissions have remained constant or declined (Newman 2017). However, the management of carbon dioxide emissions is not yet a truly poly-scalar approach. There is an absence of a clear set or rules. Targets for carbon dioxide emissions vary greatly. The Paris Agreement has a target of limiting warming to 2 °C preferably 1.5 °C. The Planetary Boundary for carbon dioxide is an atmospheric concentration of 350 ppm. The IPCC have proposed a pathway to phase out emissions by 2100 based on a global budget of CO_2 emissions. However, although the various scenarios do include bringing CO_2 back to 350 ppm which will require significant change including sucking large amounts of CO_2 from the atmosphere, they focus much more on what we can do to keep it below 450 ppm which would keep global average temperature at the 2 °C limit. This is understandable as even to get the world to begin reducing its emissions has been a 20-year programme of constant change. To get them to reduce greatly to zero and then to begin regenerating the atmosphere by various carbon removal technologies will require some very clear scientifically derived limits with associated scalar targets. A poly-scalar approach with clearly defined global targets will help increase the trust that efforts at every scale will make a difference to the end goal and that others are working towards the same end.

To enable the management of the Earth system within the PBs, results of environmental impact assessments should be compared to absolute limits rather than incremental targets. Connecting absolute limits with environmental accounting can be used to drive systemic change.

5.6   Conclusions

The science of measuring and estimating environmental impacts that have occurred or may occur in the future because of human activity is very advanced. The key limitation of these measurements is that the results are usually not able to be compared to scientific limits. In order to better manage the environmental impacts of human activity, a mechanism to compare the results of environmental impact assessments to absolute limits is needed. This is the basis of Planetary Accounting. The next chapter concludes the first section of this book by highlighting the disconnect between the findings described in Chaps. 3, 4, and 5 and introduces Planetary Accounting which has been developed to address this gap.

References

Akenji L, Bengtsson M, Bleischwitz R, Tukker A, Schandl H (2016) Ossified materialism: introduction to the special volume on absolute reductions in materials throughput and emissions. J Clean Prod 132:1

Arvidsson R, Kushnir D, Molander S, Sandén BA (2016) Energy and resource use assessment of graphene as a substitute for indium tin oxide in transparent electrodes. J Clean Prod 132:289–297

Bahadur A, Tanner T (2014) Transformational resilience thinking: putting people, power and politics at the heart of urban climate resilience. Environ Urbaniz 26:200–214

Baumann A (2013) Greenhouse gas emissions associated. Sustainable Development Masters, Uppsala University

Bellis M (2017) A history of electric vehicles. Available: https://www.thoughtco.com/history-of-electric-vehicles-1991603. Accessed 22 Apr 2018

BIS (2012) Low carbon environmental goods and services. Department for Business, Innovation and Skills, London

Biswas A, Modak P (1999) Conducting environmental impact assessment for developing countries. United Nations University, New York, NY

Cucek L, Klemes JJ, Kravanja Z (2012) A review of footprint analysis tools for monitoring impacts on sustainability. J Clean Prod 34:9–20

De Boo AJ, Bosch PR, Gorter CN, Keuning SJ (1993) An environmental module and the complete system of national accounts. In: Franz A, Stahmer C (eds) Approaches to environmental accounting. Springer, Heidelberg

Ewing B, Moore D, Goldfinger S, Ourslet A, Reed A, Wackernagel M (2010a) Ecological footprint atlas 2010. Global Footprint Network, Oakland, CA

Ewing B, Reed A, Galli A, Kitzes J, Wackernagel M (2010b) Calculation methodology for the national footprint accounts. Global Footprint Network, Oakland, CA

Fang K (2015a) Footprint family: concept, classification, theoretical framework and integrated pattern. Shengtai Xuebao/Acta Ecologica Sinica 35:1647–1659

Fang K (2015b) Footprint family: current practices, challenges and future prospects. Shengtai Xuebao/Acta Ecologica Sinica 35:7974–7986

Fang K, Heijungs R, De Snoo GR (2014) Theoretical exploration for the combination of the ecological, energy, carbon, and water footprints: overview of a footprint family. Ecol Indic 36:508

Fong WK, Sotos M, Doust M, Schultz S, Marques A, Deng-Beck C (2014) Global protocol for community-scale greenhouse gas emission inventories - an accounting and reporting standard for cities. Greenhouse gas protocol. World Resources Institute, C40 Cities, and Local Governments for Sustainability, Washington, DC

Galli A, Kitzes J, Wermer P, Wackernagel M, Niccolucci V, Tiezzi E (2007) An exploration of the mathematics behind the Ecological Footprint. Int J Ecodyn 2:250–257

Galli A, Wiedmann T, Ercin E, Knoblauch D, Ewing B, Giljum S (2012) Integrating ecological, carbon and water footprint into a "Footprint Family" of indicators: definition and role in tracking human pressure on the planet. Ecol Indic 16:100–112

Galli A, Wackernagel M, Iha K, Lazarus E (2014) Ecological footprint: implications for biodiversity. Biol Conserv 173:121–132

Gladwell M (2000) The tipping point. Little Brown, New York, NY

Global Footprint Network (2011) National footprint accounts. Global Footprint Network, Oakland, CA

Global Footprint Network (2014) Footprint calculator. Available: http://footprintnetwork.org/en/index.php/GFN/page/calculators/. Accessed 9 Jun 2014

Gray A (2017) Countries are announcing plans to phase out petrol and diesel cars. Is yours on the list?. World Economic Forum. Available: https://www.weforum.org/agenda/2017/09/countries-are-announcing-plans-to-phase-out-petrol-and-diesel-cars-is-yours-on-the-list/. Accessed 22 Apr 2018

Greenhalgh S, Broekhoff D, Daviet F, Ranganathan J, Acharya M, Corbier L, Oren K, Sundin H (2005) The GHG protocol for project accounting. Greenhouse gas protocol. World Resources Institute and World Business Council for Sustainable Development, Washington, DC

Guinée J (2012) Life cycle assessment: past, present and future. International Symposium on Life Cycle Assessment and Construction, Nantes, France

Hertwich EG (2005) Consumption and the rebound effect: an industrial ecology perspective. J Indus Ecol 9:85–98

ISO (1997) ISO 14040:1997 Environmental management - life cycle assessment - principles and framework. ISO, Geneva

Kalbar PP, Birkved M, Nygaard SE, Hauschild M (2017) Weighting and aggregation in life cycle assessment: do present aggregated single scores provide correct decision support? J Indus Ecol 21:1591–1600

Kirchain RE Jr, Gregory JR, Olivetti EA (2017) Environmental life-cycle assessment. Nat Mater 16:693–697

Kojima S, Aoki-Suzuki C (2015) Efficiency and fairness of resource use: from a planetary boundary perspective. In: The economics of green growth: new indicators for sustainable societies. CRC Press, Boca Raton, FL

Laurent A, Owsianiak M (2017) Potentials and limitations of footprints for gauging environmental sustainability. Curr Opin Environ Sustain 25:20–27

Monfreda C, Wackernagel M, Deumling D (2004) Establishing national natural capital accounts based on detailed Ecological Footprint and biological capacity assessments. Land Use Policy 21:231–246

Musk E (2014) All our patent are belong to you. Tesla. Available: https://www.tesla.com/en_NZ/blog/all-our-patent-are-belong-you. Accessed 22 Apr 2018

Newman P (2005) Can the magic of sustainability revive environmental professionalism? Greener Manag Int 49:11–23

Newman P (2017) The rise and rise of renewable cities. Renew Energ Environ Sustain 2:10

Newman P, Kenworthy J (2006) Urban design to reduce automobile dependance. Opolis 2:35–52

Newman P, Kenworthy J (2015) The end of automobile dependence: how cities are moving beyond car-based planning. Island Press, Washington, DC

Newman P, Beatley T, Boyer H (2017) Resilient cities: overcoming fossil fuel dependence. Island Press, Center for Resource Economics, Washington, DC

OVERLAET (2011) Tesla_Roadster_2.5_(fron_quarter).jpg. Wikimedia

Oxford Dictionary (2018) Systemic. Oxforddictionaries.com. Available: https://en.oxforddictionaries.com/definition/systemic. Accessed 22 Apr 2018

Pelling M, O'Brien K, Matyas D (2015) Adaptation and transformation. Clim Change 133:113–127

Rees WE (1992) Ecological footprints and appropriated carrying capacity: what urban economics leaves out. Environ Urbaniz 4:121–130

Saebo HV (1994) Natural resource accounting - the Norwegian approach. UNEP/ECE/UNSTAT Workshop on Environmental and Natural Resource Accounting, Modra-Harmonia

Sandin G, Peters GM, Svanström M (2015) Using the planetary boundaries framework for setting impact-reduction targets in LCA contexts. Int J Life Cycle Assess 20:1684–1700

Seijts GH, Gandz J (2018) Transformational change and leader character. Bus Horiz 61:239

SFOSKETT (2005) 1915_Detroit_Electric.jpg. Wikimedia

Termeer K, Dewulf A, Biesbroek R (2017) Transformational change. J Environ Plan Manag 60:558–576

UN (1993) Handbook of national accounting: integrated environmental and economic accounting. United Nations, New York, NY

UN (2014) System of environmental-economic accounting 2012 - central framework. United Nations, New York, NY

UN Statistics Division (2018) SEEA revision. United Nations Department of Economic and Social Affairs, Statistics Divisions. Available: https://unstats.un.org/unsd/envaccounting/seearev/

Valada T (2010) Ecological footprint: an indicator of environmental (un)sustainability? A review and further analysis. Instituto Superior Técnico, Lisboa

Wiedmann T (2009) A review of recent multi-region input–output models used for consumption-based emission and resource accounting. Ecol Econ 69:211–222

Chapter 6
Resolving the Disconnect Between Earth-System Science, Management Theory, and Environmental Accounting

We build too many walls and not enough bridges.
Isaac Newton

Abstract The opportunity to use the Planetary Boundaries to inform policy, behaviour, and environmental management has been highlighted by several authors. There have been several adaptations of the Planetary Boundaries: some connecting them to environmental assessment systems and others adapting them to form the basis of national or regional environmental accounts.

However, the Planetary Boundaries were not designed to be scaled or to be related to human activity. They were intended as indicators of the scale and urgency of the problem, not as a guide to resolving it. This is apparent in the adaptations, each of which has limitations pertaining to the metrics used, limits selected, further scalability, relatability to human activity, or comparability to other adaptations.

Insights derived from environmental accounting theories can be used to understand and resolve the disconnect between the Planetary Boundaries and the management of environmental impacts. The European Environment Agency Driver-Pressure-State-Impact-Response framework shows how different environmental indicators can be categorised. For indicators to scale and relate easily to human activity, the pressure category of indicators is required. The planetary boundary indicators are not of a uniform category. Some are pressures; however, most are states or impacts. State and impact indicators are not easy to scale or to relate to human activity.

In order to connect environmental accounting practices with the Planetary Boundaries in a way that enables a poly-scalar approach to Earth-system management, we have identified the need to translate the boundaries into a uniform set of pressure indicators; to enable the expansion of the concepts of carbon accounting and science-based targets across all of the Planetary Boundaries. This is the basis of Planetary Accounting.

© Springer Nature Singapore Pte Ltd. 2020
K. Meyer, P. Newman, *Planetary Accounting*,
https://doi.org/10.1007/978-981-15-1443-2_6

6.1   Introduction

The previous chapters concluded that:

- Humankind is changing the state of the planet from one that is hospitable to humans to one that is likely to become increasingly hostile to human habitation.
- The Planetary Boundaries define a safe operating space for humanity.
- It is likely that we will need systemic rather than incremental change to be able to operate within the safe operating space.
- The most effective way to manage global environmental problems is through a poly-scalar approach which is integrative across different scales and which is coordinated by a general system of rules.
- This approach will need to be underpinned by environmental accounting—an important scientific tool for decision-makers driving environmental management.
- Absolute environmental limits (such as the Planetary Boundaries) can be used to create systemic rather than incremental change as they lend themselves to processes such as disruptive innovation.

In summary, the Planetary Boundaries need to be connected to environmental accounting methods in a way that enables a poly-scalar approach to Earth-system

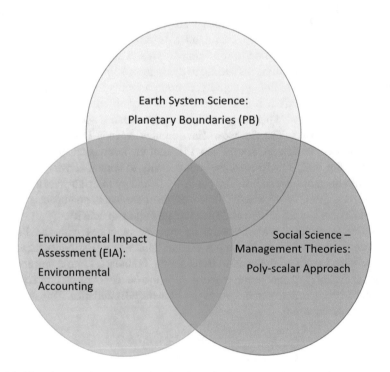

Fig. 6.1 There is a need to connect theories from Earth-system science, environmental impact assessment, and management theories

management as shown in the previous chapters. There is a need to connect theories from Earth-system science, environmental impact assessment, and management theories as shown in Fig. 6.1.

The problem is that the Planetary Boundaries in their current form cannot be connected to environmental accounting methods in a way that would enable a polyscalar approach. This chapter uses an analysis of past adaptations of the Planetary Boundaries to prove this premise. It then applies insights from environmental accounting to explain why, and to show what is needed, to translate the Planetary Boundaries into a framework which is more accessible to policymakers, technological innovators, and behaviour change at all levels.

6.2 Adapting the Planetary Boundaries

The first Planetary Boundaries' article was limited to presenting global limits (Rockström et al. 2009). One of the significant additions in the 2015 update was the inclusion of regional limits for many of the Boundaries, in acknowledgement of the importance of changes at regional and local levels on the global functioning of the Earth system (Steffen et al. 2015).

Several authors have highlighted the opportunity for the Planetary Boundaries to be used to reform environmental governance at multiple scales (e.g. Galaz et al. 2012a, b; Cole et al. 2014; Akenji et al. 2016). The Planetary Boundaries have been used as the basis for the development of national and regional environmental targets, e.g. for Switzerland, Sweden, South Africa, the Netherlands, and the European Union (Nykvist et al. 2013; Dao et al. 2015; Cole et al. 2014; Lucas and Wilting 2018; Hoff et al. 2014). There have also been several studies linking environmental footprints and life-cycle assessment (LCA) to the Planetary Boundaries (e.g. Fang et al. 2015a; Sandin et al. 2015; Chandrakumar and McLaren 2018). The latter studies typically focus on the linkages rather than using the Planetary Boundaries to derive absolute limits for an environmental accounting framework. Table 6.1 summarises the national and regional adaptations as well as an example of works linking environmental footprints and life-cycle assessments to the PBs. Each of these studies is described in more detail in Table 6.4.

Each of these adaptations is described in more detail below.

6.2.1 National Limits for Switzerland

Dao et al. (2015) converted five of the nine PB limits to national level limits. They excluded four PBs with the following justifications:

- Ozone depletion—ozone-depleting substances are being phased out under the Montreal Protocol.

Table 6.1 Adaptations of the Planetary Boundaries

Purpose	Approach	Reference
Determining national limits for Switzerland	Scaling limits using a top-down approach based on equal per capita share for past, present, and future generations	Dao et al. (2015)
Determining national limits for Sweden	Scaling limits using a top-down approach based on equal per capita share	Nykvist et al. (2013)
Determining national limits for South Africa	Bottom-up approach based on national resources	Cole et al. (2014)
Determining national limits for the Netherlands	Six different approaches considered	Lucas and Wilting (2018)
Determining regional limits for the EU	Top-down approach based on equal per capita share	Hoff et al. (2014)
Linking to environmental footprint assessments	Intermediate metrics to link top-down and bottom-up	Fang et al. (2015b)
Linking to life-cycle assessments	Relating impact categories from LCA to PB metrics	Sandin et al. (2015)

- Water—lack of evidence of a global limit.
- Aerosols and novel entities—lack of rationales from which to set limits.

The Swiss study is based on an equal per capita share for past, present, and future generations (Dao et al. 2015). It is the only adaptation of the Planetary Boundaries to include a temporal element in the derivation of a share of the safe operating space.

In the derivation of national limits, the authors of the Swiss study began by selecting new control variables for each of the Earth-system processes. Where possible, they selected indicators which describe the state of the environment. Other types of indicators were selected only when they found specific justification to deviate, for example, the lack of a suitable state indicator. They then determined global limits for the control indicators selected—noting that some of these were annual limits, while others were limits over time. Finally, they calculated the Swiss share of each global limit.

Of particular note is their limit for CO_2 emissions. The Swiss CO_2 threshold is set based on a warming limit of 2 °C. The authors acknowledge that this does not correspond to the safe operating space but suggest that this figure is more congruent with existing indicators used in Switzerland. They argue that a more stringent condition would have little effect due to the already significant challenges involved in remaining below 2 °C.

Another unusual element of the Swiss approach is their temporal allocation of CO_2 emissions. Accounting for projected population growth, they find that this gives 1.7 tCO_2e per capita, per year (until 2100), a limit for the world of 12.3 $GtCO_2e$ in 2015. Their overall budget is based on IPCC projections of a 50% chance of remaining below 2 °C. This gives an average of 15.5 $GtCO_2e$ per year until 2100. In contrast to typical pathways that show reducing CO_2 emissions over time, their methodology allows for increasing emissions with increasing population, i.e. the gross annual budget is lowest in the first year and highest in 2100.

6.2.2 National Limits for Sweden

Nykvist et al. (2013) converted four of the PBs to national limits—PBs for climate change, water, nitrogen, and land use. They omitted the PB for novel entities because of a lack of detailed and accurate data. They did not propose limits for biodiversity loss, ozone depletion, phosphorous, or novel entities but identified alternate indicators to assess these at a national level. They did not propose an indicator or limit for the PB for aerosols.

Like the Swiss limits, the Swedish limits were calculated based on an equal per capita share. However, the Swedish limits did not take into account time or population growth. Unlike the Swiss limits, the Swedish adaptation prioritises control variables pertaining to environmental flows (such as emissions of CO_2 or consumption of water) rather than environmental states.

It is interesting to compare the Swedish and Swiss limits for CO_2. The Swiss per capita limit is 1.7 tCO_2e per person per year. The Swedish limit is 2 tCO_2 per person per year. Initially these values may seem very close. However, the Swedish limit is far more lenient as it considers only emissions of CO_2, whereas the Swiss limit of CO_2 *equivalent* also includes emissions of all other GHGs (Dao et al. 2015).

6.2.3 National Limits for South Africa

Cole et al. (2014) took a very different approach to adapting the PBs to national indicators for South Africa. They assessed each PB against the following criteria:

- Is this relevant at a national scale?
- Does the set of dimensions include the main environmental and social concerns in South Africa?

Where the response to both questions was positive, they then set about determining the most appropriate indicators based on the availability of data and a means to determine a national level limit. The limits they proposed were not based on a global share of the PBs. They were determined using a bottom-up approach. Their limit for biosphere integrity is *no endangered or critically endangered ecosystems*. This is difficult to compare to the PB of a maximum extinction rate of ten species per million species per year. It is possibly a more stringent limit.

6.2.4 National Limits for the Netherlands

Lucas and Wilting (2018) do not prescribe a particular approach to deriving national limits from global limits. Rather, they explore six different approaches: grandfathering, equal per capita, cumulative equal per capita, ability to pay, development rights, and resource efficiency. From these six approaches, they are then able to determine a limit range, e.g. they estimate that the Netherlands should limit their carbon dioxide emissions to between −6.6 and 1.9 tCO_2 per capita. They have used this approach to determine limit ranges against four of the Planetary Boundaries, the boundaries for climate change, land-use change, biogeochemical flows, and biodiversity loss.

6.2.5   Regional Limits for the EU

Hoff et al. (2014) take yet another approach. They compare European footprints to the Planetary Boundaries and propose European level limits for each footprint.

In the footprint assessment, they consider:

- Material footprint
- CO_2 footprint
- Water footprint
- Land footprint
- Biodiversity footprint

The material footprint has no associated PB. It is worth considering that in order to operate within the PBs, it is likely that the global material footprint would need to shrink substantially. The CO_2 footprint is compared to the limit proposed for Sweden of 2 tCO_2 per capita (Nykvist et al. 2013) rather than the planetary boundary of atmospheric concentration of $CO_2 \leq 350$ ppm. Of the footprints assessed, only the water and land footprints were directly compared to the PBs. Although there is a PB for biodiversity, the control variable used to assess the state of biodiversity in the EU cannot be compared to the PB control variable.

6.2.6   The FB-ESA Framework

Fang et al. (2015a) proposed a framework for the conversion of the Planetary Boundaries to footprint indicators, the "Footprint-Boundary Framework for Environmental Sustainability Assessment" (FB-ESA). The authors converted the control variables of the boundaries, and control variables of selected footprint variables, to create a set of common control variables. These control variables allow the results of bottom-up footprint assessments to the global limits defined by the PBs.

The FB-ESA framework is the only adaptation that considers the interconnectivity of the PBs. They identify that this overlapping of multiple footprint and PBs is a weakness for the FB-ESA. They do not propose a mechanism to deal with this interconnectivity. The authors have applied their framework in a review of 28 countries (Fang et al. 2015b). This study was limited to the assessments of the carbon footprint, water footprint, and land footprint of each country.

6.2.7 Connecting the PBs and LCA

Sandin et al. (2015) proposed life-cycle assessment (LCA) impact categories which could be related to the PB control variables. They use this connectivity to propose impact reduction targets at the scale of products.

This is an interesting study as its primary focus is on products (as opposed to groups of people). They use a distance-to-target method to translate the limits from the PB variables to the impact category variables. This means they compare the percentage overshoot of impacts to each limit.

Such an approach does not scale sensibly. At a global scale, it is reasonable to assume that an 80% overshoot of the PBs for nitrogen implies that 80% reductions are needed for the LCA impact category "eutrophication"—the impact of nitrogen consumption. However, it does not translate into finite targets at different scales. It would not make sense for all sectors to target 80% reductions in eutrophication. Consider two producers of beans: One may use twice as much nitrogen per kg of beans than the other. It might be quite easy for producers who are using twice as much nitrogen to reduce their usage by 80%. For the producer who already uses nitrogen sparingly, this reduction may not be feasible. Moreover, care should be taken in the conclusions drawn from the distance to target. A planetary boundary with a very high distance to target does not necessarily pose more risk than a planetary boundary with a low distance to target.

6.2.8 Comparing the Adaptations of the Planetary Boundaries

Table 6.2 shows the limits and/or control variables proposed in the adaptations listed above against the Planetary Boundaries. A comparative analysis of each row of this table shows wide variance between the indicators and the proposed limits.

The adaptations begin to connect Earth-system science with environmental accounting as shown in Fig. 6.2. However, none of these provide a suitable basis for a poly-scalar approach to Earth-system management. This is because they cannot easily be applied to scales other than the scale for which each is intended. For example, very few of the indicators use to assess biodiversity impacts could be used at the scale of a business, an individual, a product, or a service.

Environmental footprint and life-cycle assessments are generally suited to a poly-scalar approach (see Fig. 6.2). The studies that link the PBs to environmental impact assessment mechanisms were developed to enable the comparison of impacts of different types of activity to global limits. Yet even these have very limited scalability. For example, the F-B-ESD framework proposed a water footprint limit of 40% of the total renewable water resources for that country. It may be possible to apply a similar approach to a regional or city scale. However, it would not make sense at a business or an individual scale. The biogeochemical flow indicator proposed for the LCA adaptation is eutrophication (Sandin et al. 2015). In some instances, this may be attributable to a single source and thus be scalable. In many cases, it would not be. It is not an indicator that makes sense consistently at different scales.

Further, the level of work that has gone into each of the adaptations is extremely high. It would not be feasible to apply this much effort for every application of the Planetary Boundaries to different decision-making centres.

Moreover, even the connectivity between the adaptations and the Planetary Boundaries is limited. For example, the PB for climate change has two distinct limits:

1. The concentration of carbon dioxide in the atmosphere should not exceed 350 parts per million (ppm) (parts of atmosphere).
2. The change in radiative forcing (the energy balance at Earth's surface) should not exceed ± 1 W/m^2.

Both of these limits have been exceeded. To return to 350 ppm would mean removing CO_2 from the atmosphere. Yet in the adaptations, all the CO_2 budgets except the lower end of the range for the Netherlands are positive, i.e. they allow for ongoing CO_2 emissions. Following any of these targets would thus lead to further overshoot of the PB for CO_2 concentration. Further, the indicators used vary widely. Some include all greenhouse gases. Others are limited to CO_2. None of the adaptations address the PB limit for radiative forcing.

Another example of high variability is the adaptations for biosphere integrity. The PB for biosphere integrity is a global extinction rate of no more than ten extinct species per million species per year (E/MSY). The control variables in each adaptation are shown in Table 6.3. Biosphere integrity is one of the PBs that has already been exceeded. It is thus one of the most critical limits. However, each of the control variables used in the adaptations is different. It is not straightforward to assess whether the limits in each adaptation are equivalent to the PB for biosphere integrity or not.

Finally, several of the PBs are excluded from some or all of the adaptations, for example, the PB for atmospheric aerosol density—*aerosol optical depth*.

It could be argued that in a poly-scalar order, it does not matter whether the indicators and exact limits are the same or not as long, as they all target roughly the same end goal. However, as discussed in Chap. 4, one of the most important elements for a poly-scalar approach to be effective is that it creates a sense of trust that others are making a similar effort. The past adaptations of the PBs are not consistent with one another or with the Planetary Boundaries themselves. As such, it is unlikely that these would generate substantial trust between different stakeholders.

As portrayed in Fig. 6.2, there is a disconnect between the Planetary Boundaries, environmental accounting, and a poly-scalar approach. This must be resolved if we are to create true and useful Planetary Accounting.

Table 6.2 Planetary Boundaries downscaled for use at sub-global levels

Earth-system process (Rockström et al. 2009)	Planetary boundary threshold	Adapted boundary for Sweden (per capita) (Nykvist et al. 2013)	Adapted boundary for Switzerland (global limits) (Dao et al. 2015)	Adapted boundary for South Africa (national limits) (Cole et al. 2014)	Adapted boundary for the Netherlands (per capita limit ranges) (Lucas and Wilting 2018)	Adapted boundary based on the FB-ESA framework (per capita) (Fang et al. 2015b)	Corresponding LCA impact category (Sandin et al. 2015)
Climate change	Atmospheric CO_2	CO_2 emissions: 2 tCO_2/capita/year	CO_2 emissions: 12.3 $GtCO_2$eq	CO_2 emissions: 451 $MtCO_2$	CO_2 emissions: −6.6 to 1.9 tCO_2/capita/year	Carbon footprint: 3.1 tCO_2-eq/year	Climate change
	Change in radiative forcing	NA	NA	NA	NA	NA	NA
Biodiversity loss	10 species/million species extinct per year	*No boundary set; however, three alternative indicators identified: (number of species threatened within the national territory, number of species threatened globally, percentage of marine and terrestrial areas protected)*	Biodiversity damage potential: 0.16	Endangered and critically endangered ecosystems: 0%	Land-use change: 0.1–0.9 ha/capita/year	NA	Land occupation, land transformation, biodiversity loss

(continued)

Table 6.2 (continued)

Earth-system process (Rockström et al. 2009)	Planetary boundary threshold	Adapted boundary for Sweden (per capita) (Nykvist et al. 2013)	Adapted boundary for Switzerland (global limits) (Dao et al. 2015)	Adapted boundary for South Africa (national limits) (Cole et al. 2014)	Adapted boundary for the Netherlands (per capita limit ranges) (Lucas and Wilting 2018)	Adapted boundary based on the FB-ESA framework (per capita) (Fang et al. 2015b)	Corresponding LCA impact category (Sandin et al. 2015)
Nitrogen and phosphorus cycle	N_2 removed from the atmosphere	Nitrogen emissions: 5 kg/capita/year	Nitrogen emissions: 47.6 Tg	Nitrogen application rate for maize production: 144 kg N/ha	Intentional nitrogen fixation: −10.8 to 19.3 kg N/capita/year	NA	Eutrophication: marine eutrophication, terrestrial eutrophication, terrestrial acidification
	P flowing into oceans	No boundary set; however, alternative indicator identified (phosphorous fertiliser consumption)	Phosphorous fertiliser consumption: 38.5 Tg (global limit)	Total phosphorous concentration in dams: 0.1 mg/L	P fertiliser use: −3.7 to 2.2 kg P/capita/year		Eutrophication: freshwater eutrophication
Stratospheric ozone depletion	Concentration of ozone	No boundary set; however, alternative indicator identified (ozone-depleting potential)	Not considered as currently phased out	Annual HCFC consumption: 369.7 ODPt	NA	NA	NA
Ocean acidification	Mean saturation state of aragonite: 2.75	Not assessed as ocean acidification is an impact of climate change	CO_2 emissions: $GtCO_2$ 7.6 $GtCO_2$	Replaced by marine harvesting	NA	NA	NA

Earth-system process (Rockström et al. 2009)	Planetary boundary threshold	Adapted boundary for Sweden (per capita) (Nykvist et al. 2013)	Adapted boundary for Switzerland (global limits) (Dao et al. 2015)	Adapted boundary for South Africa (national limits) (Cole et al. 2014)	Adapted boundary for the Netherlands (per capita limit ranges) (Lucas and Wilting 2018)	Adapted boundary based on the FB-ESA framework (per capita) (Fang et al. 2015b)	Corresponding LCA impact category (Sandin et al. 2015)
Freshwater use	Freshwater consumption: 4000 km³/year	Water consumption: 585 m³/capita/year	Not included as considered to be a regional issue	Consumption of available freshwater resources: 14,196 Mm³/year	NA	Water footprint: 40% of the total renewable water resources for that country	Freshwater consumption
Change in land use	Land cover converted to cropland: 15%	0.3 ha/capita	Surface of anthropised land: 19,362,000 km²	Rain-fed arable land converted to cropland: 12.1%	Cropland Use: 0.1–0.5 ha/capita/year	Land footprint—biocapacity	Land transformation
Chemical pollution	NA	*No boundary set; however, 5 alternative indicators identified (pesticide regulation, persistent organic pollutants in breastmilk, methylmercury-based indicator, embedded use of chemical substances in traded products, use of the Strategic Approach to International Chemicals Management)*	Rationales lacking in setting limit	Not given due to lack of detailed and accurate data	NA	NA	NA

(continued)

Table 6.2 (continued)

Earth-system process (Rockström et al. 2009)	Planetary boundary threshold	Adapted boundary for Sweden (per capita) (Nykvist et al. 2013)	Adapted boundary for Switzerland (global limits) (Dao et al. 2015)	Adapted boundary for South Africa (national limits) (Cole et al. 2014)	Adapted boundary for the Netherlands (per capita limit ranges) (Lucas and Wilting 2018)	Adapted boundary based on the FB-ESA framework (per capita) (Fang et al. 2015b)	Corresponding LCA impact category (Sandin et al. 2015)
Atmospheric aerosol loading	Aerosol optical depth	*No boundary set, no indicator proposed*	*Rationales lacking in setting limit*	*Replaced by air pollution*	NA	NA	NA

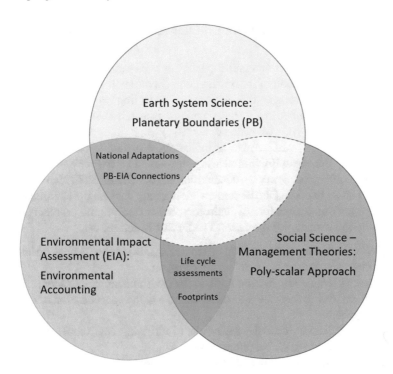

Fig. 6.2 Identifying the disconnect between Earth-system science, management theory, and environmental accounting

Table 6.3 A summary of the control variables used to assess biodiversity loss in different adaptations of the Planetary Boundaries

Adaptation	Control variable proposed for biodiversity loss
Sweden	None set but three potential control variables identified: • Number of species threatened within the national territory • Number of species threatened globally • Percentage of marine and terrestrial areas protected
Switzerland	Biodiversity damage potential—an estimation of species richness compared to background levels
South Africa	Endangered and critically endangered ecosystems: 0%
Netherlands	Land-use change
FB-ESA	None proposed
LCA-PB	• Land occupation • Land transformation • Biodiversity loss

6.3   Why the Planetary Boundaries Are Difficult to Scale

The global uptake of environmental impact assessments and environmental accounting led to the development of increasing numbers of environmental indicators with which to measure impacts on the environment. Selecting appropriate indicators for different assessments became a major topic of research in itself (e.g. Nolte et al. 2013; Dafforn et al. 2012; Chevalier et al. 2011).

In response to this vast number of indicators, a system to categorise these was adopted by the European Environment Agency—the Driver-Pressure-State-Impact-Response (DPSIR) framework—detailed in Fig. 6.3 (EEA 2005; Dao et al. 2015; Nykvist et al. 2013). The DPSIR framework not only enables the classification and therefore better understanding of indicators, but it resolves the problems outlined above with the Planetary Boundaries. The DPSIR is defined as:

- *Driver* indicators describe human needs, e.g. kilowatt hours of electricity, kilometres travelled, or litres of fuel for transport.
- *Pressure* indicators describe flows to the environment, e.g. CO_2 emissions.
- *State* indicators describe the environment, e.g. the concentration of CO_2 in the atmosphere.
- *Impact* indicators describe the results of changing environmental states, e.g. change in average global temperature.
- *Response* indicators describe human responses to the environment which can target any other level of indicator, e.g. the Paris Agreement.

To understand why the PBs are so difficult to scale or to connect to existing environmental impact assessment frameworks, it is helpful to consider the DPSIR category of each PB indicator (see Table 6.4). There are three *pressure* indicators, five *state* indicators, and one *impact* indicator.

States and *impacts* are important indicators as they communicate the status quo. However, human activity cannot easily be compared to *states* or *impacts*. This is because human activity directly influences *drivers* and *pressures*, but only indirectly influences *states* and *impacts*. *States* and *impacts* are also inherently difficult to scale. It should be noted that while most pressure indicators can be easily related to human activity, some cannot (see Box 6.1).

Take one of the PB indicators for climate change as an example—the concentration of CO_2 in the atmosphere—a *state* indicator. There is no straightforward way to divide the responsibility of the state of the atmosphere between different nations, cities, regions, or individuals. Nor can one compare specific human activities to this unit of measure. Imagine an individual deciding whether to take the car or the train to work, or a local government deciding whether to proceed with certain infrastructure—neither could begin to estimate the impacts of these decisions on the atmospheric concentration of CO_2. In contrast, CO_2 emissions, a *pressure* indicator, in this case CO_2 emissions, can easily be used as the basis for creating budgets at different scales and as the basis for decision-making.

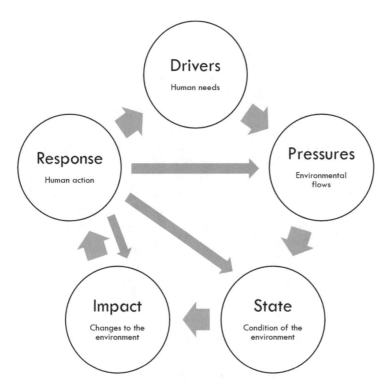

Fig. 6.3 The Driver-Pressure-State-Impact-Response is a framework that can be used to categorise environmental indicators

Table 6.4 The Planetary Boundaries in the DPSIR framework

Earth-system process	Indicator	DPSIR category
Climate change	Atmospheric concentration of CO_2	State
	Change in radiative forcing	State
Biodiversity loss	Extinction rate	Impact
Nitrogen and phosphorus cycle	N_2 removed from the atmosphere	Pressure
	P flowing into oceans	Pressure
Stratospheric ozone depletion	Atmospheric concentration of ozone	State
Ocean acidification	Mean saturation state of aragonite in the oceans	State
Freshwater use	Freshwater consumption	Pressure
Change in land use	Percentage of land cover converted to cropland	State
Novel entities	NA	NA
Atmospheric aerosol loading	Aerosol optical depth	State

Seven of the ten control variables for the Planetary Boundaries are either states or impacts. These control variables thus limit the Planetary Boundaries' scalability and relevance to decision-making.

> **Box 6.1 Different Pressure Indicators**
> Not all pressure indicators can be directly related to human activity. Some of the PBs are already *pressures*, for example, freshwater use. It is straightforward to determine the amount of freshwater used for a given activity and then to compare this to the global limit for freshwater use. Other *pressure* indicators cannot be so easily compared to human activity. For example, the PB for phosphorous is the amount of phosphorous flowing into the oceans. Humans mine phosphorous for a number of uses, the main being fertilisers. The pathway of phosphorous from the application of fertiliser to a downstream flow of phosphorous into the oceans is complex. There are many human and natural factors which influence how much phosphorous travels to the ocean, how quickly, and in which ways. It would be difficult to relate this indicator directly to human activity, even though it is a pressure indicator.

6.4   Conclusions: A Need for New Global Limits

Part I of this book has demonstrated the need to manage human impacts on the Earth system such that we can return to and live within the Planetary Boundaries. We have argued that the most effective way to do this is by connecting environmental accounting to absolute environmental limits (the Planetary Boundaries) in a way that enables a poly-scalar approach to global environmental management. However, the PB limits are not in suitable control variables. Past adaptations of the Planetary Boundaries and environmental accounting theory (the DPSIR) framework were used to explain the limitations of the PBs for such an approach.

What is needed is a new set of global limits that are communicated in terms of pressure indicators that can be directly compared to any scale or type of human activity. Such limits would enable a new type of environmental accounting where impacts can be compared to scientific, absolute limits at any scale.

This is the basis of Planetary Accounting—which is explained in the pursuing sections of this book.

References

Akenji L, Bengtsson M, Bleischwitz R, Tukker A, Schandl H (2016) Ossified materialism: introduction to the special volume on absolute reductions in materials throughput and emissions. J Clean Prod 132:1–12

Chandrakumar C, McLaren SJ (2018) Exploring the linkages between the environmental sustainable development goals and planetary boundaries using the DPSIR impact pathway framework. In: Benetto E, Gericke K, Guiton M (eds) Designing sustainable technologies, products and policies: from science to innovation. Springer International Publishing, Cham

Chevalier B, Reyes-Carrillo T, Laratte B (2011) Methodology for choosing life cycle impact assessment sector-specific indicators. ICED 11—18th International conference on engineering design—impacting society through engineering design, pp 312–323

Cole MJ, Bailey RM, New MG (2014) Tracking sustainable development with a national barometer for South Africa using a downscaled "safe and just space" framework. Proc Natl Acad Sci U S A 111:E4399–E4408

Dafforn KA, Simpson SL, Kelaher BP, Clark GF, Komyakova V, Wong CKC, Johnston EL (2012) The challenge of choosing environmental indicators of anthropogenic impacts in estuaries. Environ Pollut 163:207–217

Dao H, Peduzzi P, Chatenoux B, De Bono A, Schwarzer S, Friot D (2015) Environmental limits and Swiss footprints based on planetary boundaries. UNEP/GRID-Geneva & University of Geneva, Geneva

EEA (2005) EEA core set of indicators. Guide. EEA, Luxembourg

Fang K, Heijungs R, De Snoo GR (2015a) Understanding the complementary linkages between environmental footprints and planetary boundaries in a footprint-boundary environmental sustainability assessment framework. Ecol Econ 114:218–226

Fang K, Heijungs R, Duan Z, De Snoo GR (2015b) The environmental sustainability of nations: benchmarking the carbon, water and land footprints against allocated planetary boundaries. Sustainability (Switzerland) 7:11285–11305

Galaz V, Biermann F, Crona B, Loorbach D, Folke C, Olsson P, Nilsson M, Allouche J, Persson Å, Reischl G (2012a) 'Planetary boundaries'—exploring the challenges for global environmental governance. Curr Opin Environ Sustain 4:80–87

Galaz V, Crona B, Österblom H, Olsson P, Folke C (2012b) Polycentric systems and interacting planetary boundaries - emerging governance of climate change-ocean acidification-marine biodiversity. Ecol Econ 81:21–32

Hoff H, Nykvist B, Carson M (2014) "Living well, within the limits of our planet"? Measuring Europe's growing external footprint. SEI Working Paper 2014–05. Stockholm Environment Institute, Sweden

Lucas P, Wilting H (2018) Towards a safe operating space for the Netherlands: using planetary boundaries to support national implementation of environment-related SDGs

Nolte C, Agrawal A, Barreto P (2013) Setting priorities to avoid deforestation in Amazon protected areas: are we choosing the right indicators? Environ Res Lett 8

Nykvist B, Persson Å, Moberg F, Persson LM, Cornell SE, Rockström J (2013) National environmental performance on planetary boundaries: a study for the Swedish Environmental Protection Agency. Swedish Environmental Protection Agency, Sweden

Rockström J, Steffen W, Noone K, Persson A, Chapin FS III, Lambin E, Lenton TM, Scheffer M, Folke C, Schellnhuber HJ, Nykvist B, De Wit CA, Hughes T, van der Leeuw S, Rodhe H, Sorlin S, Snyder PK, Costanza R, Svedin U, Falkenmark M, Karlberg L, Corell RW, Fabry VJ, Hansen J, Walker B, Liverman D, Richardson K, Crutzen P, Foley J (2009) Planetary boundaries: exploring the safe operating space for humanity. Ecol Soc 14:32

Sandin G, Peters G, Svanström M (2015) Using the planetary boundaries framework for setting impact-reduction targets in LCA contexts. Int J Life Cycle Assess 20(12):1684–1700

Steffen W, Richardson K, Rockström J, Cornell SE, Fetzer I, Bennett EM, Biggs R, Carpenter SR, De Vries W, De Wit CA, Folke C, Gerten D, Heinke J, Mace GM, Persson LM, Ramanathan V, Reyers B, Sörlin S (2015) Planetary boundaries: guiding human development on a changing planet. Science 347:1259855

Part II
Developing Planetary Quotas

Chapter 7
Translating the Planetary Boundaries into Planetary Quotas

Science is simply the word we use to describe a method of organising our curiosity
Tim Minchin

Abstract There is a need to translate the Planetary Boundaries into a new set of global limits that are in *Pressure* indicators (rather than *States* or *Impacts*), which can be scaled and compared directly to the environmental impacts of human activity. We call these new Planetary Boundary-translated limits "Planetary Quotas". Planetary Quotas quantify what we need to do to return to and operate within the Planetary Boundaries. They provide the basis for a transparent and scientific approach to Planetary Accounting.

The scientific methodologies used to derive the Planetary Quotas are based on post-normal science and integrative thinking. To address the risk of unacceptable levels of normative judgements in this project, we used extended-peer community engagement, i.e. we sought feedback from a wide range of international experts in relevant fields of research.

The European Driver-Pressure-State-Impact-Response (DPSIR) framework formed the basis of our research methods. We advanced past translations of the Planetary Boundaries using the DPSIR framework by translating the Planetary Boundaries collectively rather than individually. This allowed us to better capture the interconnectivity of the Planetary Boundaries within the Planetary Quotas. Further, our methods extended across all of the Planetary Boundaries where past works have only translated some.

The results are a set of ten Planetary Quotas. The Planetary Quotas enable a new approach to global environmental management which we call "Planetary Accounting". Planetary Accounting describes the process of comparing the environmental impacts of different scales of human activity to a share of global environmental limits—the Planetary Quotas. This would allow meaningful decisions to be made at various scales on regulating activities, urban planning, design and technology, policy, industry, and all levels of government legislation.

© Springer Nature Singapore Pte Ltd. 2020
K. Meyer, P. Newman, *Planetary Accounting*,
https://doi.org/10.1007/978-981-15-1443-2_7

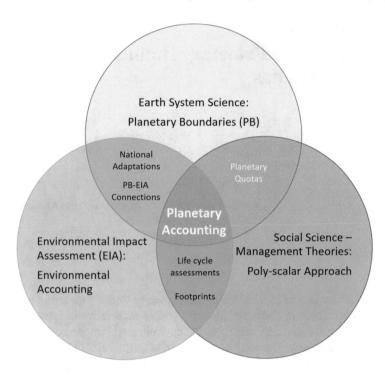

Fig. 7.1 Planetary Quotas are needed to enable Planetary Accounting, a new approach to managing the global environment

The Planetary Quotas and Planetary Accounting enable the practical application and communication of how to live within the Planetary Boundaries at different scales of human activity.

7.1   Introduction

Part I of this book concludes that there is a need to connect the Planetary Boundaries with environmental accounting practices in a way that enables a poly-scalar approach to Earth-system management. In order to do this, we identified the need for a new set of global environmental limits which we call Planetary Quotas. Planetary Quotas enable a new approach to managing global environmental problems—which we call Planetary Accounting—and which is set out in Fig. 7.1.

Planetary Accounting is the process of comparing the environmental impacts of human activity at different scales to a corresponding share of global, scientifically determined limits for human pressures on the environment—the Planetary Quotas. The Planetary Quotas are derived from the Planetary Boundaries. Each is in a

Pressure indicator that can be scaled and directly compared to the impacts of human activity.[1]

The purpose of the Planetary Quotas is to complement rather than replace the Planetary Boundaries. The Planetary Boundaries are the parameters for a planetary state that is favourable to humanity. Measuring the current situation against these parameters provides important information about the state of the global environment and the level of risk that human activity will fundamentally change this. For example, the PBs tell us that the concentration of CO_2 in the atmosphere is too high, that the current rate of species loss is dangerous, and that the acidity of the oceans is putting the Earth system at risk. In this way, the Planetary Boundaries can be likened to a planetary health check. If a person visits the doctor, she might measure his blood pressure, heart rate, weight, and liver and heart function. By comparing these results to scientifically determined healthy limits, she can determine whether he is in good health or not and, if not, which areas of his health need the most focus.

Health checks are important. They can give an early indication that there are warning signs for major health problems. Is the patient likely to experience liver failure? Is he at risk of a heart attack? However, these indicators of the state of the patient's health do not tell the patient what to do. For this reason, the doctor would almost certainly tell her patient what he should do to return to good health. She might prescribe a minimum level of exercise and a maximum calorific intake. She may even suggest that some behaviours are unacceptable—for example, she is likely to say that no amount of smoking is safe.

Where the Planetary Boundaries can be likened to a health check, the Planetary Quotas can be likened to a prescription for good health. They show how to return to and operate within the Planetary Boundaries—the parameters for a healthy planet. For example, the Planetary Quotas define limits for emissions of CO_2 to the atmosphere, minimum reforestation rates, and maximum land occupation that would facilitate a return to the safe operating space.

In the same way that an overweight patient who begins a diet will not immediately return to good health, adhering to the PQs will not immediately return the Earth-system processes to within the PBs. However, with time, the health of the patient and the planet will gradually improve and eventually return to the recommended levels. Ultimately we can hope to return to the "safe operating space" within all the Planetary Boundaries as translated into Planetary Quotas.

The Planetary Quotas also help us to understand the direction of travel with respect to the safe operating space of the PBs. For example, one could estimate that the health of a currently healthy patient with very unhealthy habits is likely to deteriorate. The current state of the planet is still within some of the Planetary Boundaries. However, if we were to exceed the PQs that correspond to these PBs each year, then we could predict with reasonable confidence that over time we would exceed these PBs.

---

[1] Environmental indicators can be classified as Drivers, Pressures, States, Impacts, or Responses. See Chap. 6 for further information.

Planetary Accounting thus enables the practical application and communication of how to live within the Planetary Boundaries at different scales of human activity. Understanding human impacts in the context of global limits would allow meaningful decisions to be made at various scales on regulating activities, urban planning, design and technology, policy, industry, and different levels of government legislation.

This chapter describes the methodology and methods used to arrive at the solution of Planetary Accounting and to derive the Planetary Quotas. A more detailed account of the methodologies and methods described here is available in the PhD by Meyer (2018).

7.2   Methodologies for Multidisciplinary Research

There is no singular field of science that encompasses the research and solutions presented in this book. The Planetary Quotas and Planetary Accounting Framework are solutions derived from scientific investigation across the broad fields of science, governance, and community. There are very few methodological frameworks for research that deal with the intersection between these fields (Brandt et al. 2013; Stocker and Burke 2017). Multidisciplinary research design is often based on a combination of different methodologies from the relevant disciplines.

Traditionally, scientific investigation was assumed to have a high degree of certainty and to be free of value judgements. Due to the urgency of environmental problems facing society today, this is changing. Science with limited certainty needs to be used to make decisions as to how best to manage the environment.

7.2.1   Post-normal Science

Funtowicz and Ravetz (1993) introduced a new concept to describe a type of research where value judgements cannot be avoided—*post-normal science*. They define post-normal science as the approach needed when decision stakes are high, systems uncertain, and decisions urgent—descriptions well suited to the management of the global environment.

One of the major risks identified for post-normal science is that in the presence of uncertain systems, values held by the agent will influence the outcomes. However Ravetz (1999) argues that this is both natural and legitimate. He makes the case that a degree of value judgement is present in all research (e.g. in balancing selectivity and sensitivity or through the decisions for manging outlier data). Value judgements in themselves do not reduce the validity of research. They are only a problem when they are not made transparent and where adequate quality-control methods are not employed. Ravetz (1999) describes the commonalities for such an approach to using science to inform governance. These include being critical and reflexive, being aware of uncertainties, the inclusion of normative perspectives, and having a strong focus on quality.

Traditional scientific approaches use peer review as a quality-control method. In a post-normal science approach, the common quality-control method is to use "extended peer communities". Extended peer communities have been used for some time to guide the development of projects and policies, even if the term has not been used. For example, when implementing new infrastructure that will affect many players, governments may invite community feedback to understand any problems and/or opportunities that they may not otherwise have understood.

The common element of extended peer community reviews is that the participants are invited to review a proposal based on science and that their feedback is used to improve the quality of the final solution. The process generally follows that a period of scientific investigation is undertaken to develop a preliminary solution. This is followed by engagement with the extended peer community. The feedback gathered from the investigation is then used to refine the solution.

Aristotle had a similar approach when he talked about different levels of scientific understanding: "first road thinking" which was based on logic and only applied to problems where simple data could be used and, "second road thinking" which was where differences of opinion were inevitable and rhetoric or group discussions were required to resolve the problem.

7.2.2 Integrative Research

Integrative research is another multidisciplinary approach that also probably goes back to antiquity but has been defined and developed in 1986 (Douglas 2005). Douglas defines this approach as a process of integrating "intuition, reason, and imagination" to address a complex problem. The underlying premise of integrative research is that opposing views or models should not be assessed to determine which is better. Rather, they should be assessed, analysed, and considered, with a view to developing a new approach that takes the best aspects from both models. Integrative research uses a systemic approach—considering the problem and solution as a whole, rather than assessing and resolving the problem in parts.

7.2.3 Methodology for the Development of the Planetary Quotas and Planetary Accounting Framework

The methodological approach we chose combines the post-normal science and integrative research approaches. An integrative approach was applied throughout the project. For example, we undertook a thorough and broad investigation of different types of environmental accounting. This allowed us to gain the insights needed to arrive at the concept of Planetary Accounting. But to apply this approach to the Planetary Boundaries is not straightforward as each of the PBs comes from thousands of scientific experiments and has been pooled and processed by a group who alone know all the details about how they were derived before publishing them (Rockström et al. 2009; Steffen et al. 2015). It required a visit to meet with this group for extended discussions.

Extended peer community engagement was therefore a major component of this project. At the early stages of the project, our engagement was organic rather than formal, and predominantly local, working with peers and friends. In the later stages of the project, we took the preliminary concept out to an international audience. Meyer connected with international experts from diverse fields including Earth-system science, Planetary Boundaries, climate change, water, nitrogen, land use, aerosols, life-cycle assessment, Environmental Footprints, global governance, resilience, and the circular economy. She presented our proposal and invited open and frank feedback.

The outcome was highly effective (albeit unexpectedly challenging; see Box 7.1). The dialogue during these discussions ranged from specific concerns over particular Quotas, to high-level discussions over the ethics of such an approach. The feedback fed into the refinement of our framework—giving it greater breadth, depth, transparency, and robustness.

7.3   Methods

The underlying framework used to derive the Planetary Quotas from the Planetary Boundaries is the European Union's Driver-Pressure-State-Impact-Response (DPSIR) framework. In Chap. 6, we introduced the DPSIR framework and showed how it could be used to classify and thus better understand the utility and purpose of different metrics. Here we show how it can be used to translate metrics from one classification to another.

There is a causal relationship between the categories of the DPSIR framework as follows:

Human needs (*Drivers*) lead to environmental flows (*Pressures*) which determine the *State*. Changing the State of the environment can lead to environmental *Impacts*.

As an example, a human need for heating can be measured in <u>kilowatts of energy</u> (the *Driver*). Creating this energy leads to an environmental <u>flow of carbon dioxide to the atmosphere</u> (the *Pressure*). The amount of carbon dioxide released determines the <u>concentration of carbon dioxide in the atmosphere</u> (the *State*). As the concentration of carbon dioxide in the atmosphere changes, this leads to a <u>change in global temperature</u> (the *Impact*). In this way, the DPSIR can be used to translate Planetary Boundary metrics into pressures metrics and thus form the basis of Planetary Accounting (Fig. 7.2). The decision to use *Pressure* rather than *Driver* indicators is discussed in Chap. 6 and in Box 7.2.

Box 7.1 The European Research Trip

To complete the extended community engagement required that Meyer spent a substantial portion of the last year of the research in Europe as this is where the majority of global experts in the field are based. She therefore packed her two preschoolers and husband into a caravan and the four of them travelled from one institute to another across Europe. Between visits to research centres, her office changed daily—from beachside cafes where she could see the kids and husband building sandcastles, to busy street side cafes in Portugal, to the corner of the caravan when no suitable alternative was available.

There were some unexpected events along the way. The car broke down at least a dozen times. There were several injuries (including the dislocation of both of Meyer's knees in a single ski accident which resulted in her first ride in an ambulance and left her to complete the final month in Europe in two full leg braces). The climax was the discovery that her family's by then beloved caravan had been stolen 10 years before. The Stuttgart police relieved Meyer and her family of their home with 30 min notice, leaving them stranded as the sun was going down, with a handful of quickly gathered essentials, a broken-down car, two leg braces, and no accommodation. This and other highlights and lowlights of the trip are documented in Meyer's travel blog (Meyer 2017).

Despite the hiccups, Meyer maintains that the research trip was an incredible journey both personally and professionally. On the extended peer review, she met so many inspiring experts across the different fields of this research project and built relationships that will last for many years to come. It was also a remarkable experience for her and her family on a personal level. Her now 4-year-old daughter can still remember words in at least five languages 18 months after leaving Europe. Her son, now 6, is finally coming to terms with why the view from the window of their home stays the same every day. The whole family speak wistfully of the caravan year and are already making plans for the next sabbatical.

7.3.1 Interconnectivity

One of the key challenges in such an approach is the interconnectivity of Earth-system processes. Many of the pressures associated with one Planetary Boundary also affect others. For example, consider the Planetary Boundary for biosphere integrity—global extinction rate. Species extinctions are caused by a multitude of human impacts, including climate change. Climate change has two Planetary Boundary metrics: the concentration of CO_2 in the atmosphere and the change in radiative forcing. These metrics can be translated into carbon dioxide and other greenhouse gas emissions, change in surface albedo (i.e. change in the reflectivity of Earth's surface due to change in land type/use), and aerosol density. Each of these pressures not only affects climate change; they also influence biosphere integrity. Several of the pressures

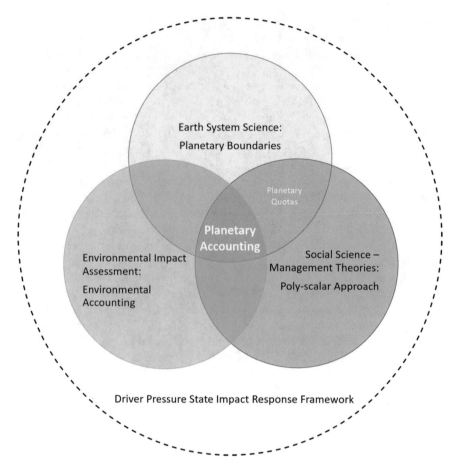

Fig. 7.2 The DPSIR framework can be used to derive Planetary Quotas and thus enables Planetary Accounting

also affect other Planetary Boundaries, e.g. for land-use change and aerosol optical depth.

Again, the example of human health can be used to describe the importance of this interconnectivity. If a person is overweight, this influences other elements of his health. His weight, blood pressure, heart rate, and liver function are not separate and independent of one another; they are all interconnected. When his doctor prescribes a diet, or exercise, she is not just targeting weight reduction. She is also aiming to improve his blood pressure, heart function, and other elements of his health.

A simple one-to-one translation of the Planetary Boundary metrics to pressure metrics would therefore have limited efficacy as they are interconnected. For the Planetary Quotas to be a robust prescription for a healthy planet, the interconnectivity of the Planetary Boundaries needs to be carried over through the translation. One could not accurately determine a Quota for global aerosol emissions based on the

Box 7.2 Why Use Pressures Not Drivers?

As discussed in Chap. 6, this was the approach used in two of the adaptations of the PBs to determine national targets for Sweden and Switzerland (Dao et al. 2015; Nykvist et al. 2013).

Both Driver and Pressure indicators can be directly applied to human activity. It is common practice, for example, to measure the carbon emissions (a Pressure) for a given activity or to monitor electricity consumption (a Driver).

The decision to use Pressure indicators rather than Driver indicators was for simplicity—using the example of atmospheric CO_2, there is only one environmental flow (Pressure) that affects this—the emission (or uptake) of CO_2 to (from) the atmosphere. In contrast, there are many human drivers (needs) which lead to the emission of CO_2. At a high level, these would include a need for transport, for electricity, for concrete, and for deforestation for agriculture. However, each of these Drivers can be traced back to underlying drivers—for example, the need for transport is due to the need to get to work and the need to socialize, and so each of these drivers is highly interconnected. Pressure is a far simpler indicator.

Planetary Boundary for aerosol optical depth without considering the impacts of this limit on climate change and biosphere integrity.

To address this interconnectivity, we applied the DPSIR translation across the Planetary Boundaries as a whole. This generated a list of "critical human pressures"—flows to the environment, caused by human activity, that affect the Planetary Boundary Earth-system processes. This list of critical pressures was distilled into a concise set of Planetary Quota metrics. The Quotas were then derived from all of the Planetary Boundaries corresponding to that metric.[2]

7.4 From PBs to PQs

The linkages between the Planetary Boundaries, critical pressures, and Planetary Quotas are shown in Fig. 7.3. The Earth system—the fundamental core for both the Planetary Boundaries and Planetary Quotas—is shown in the centre. The causal relationships can be traced from each of the Planetary Quotas, through critical pressures and then one or more Planetary Boundaries—showing how human activity affects the Planetary Boundaries and therefore the overall health of the Earth system.

The purpose of Fig. 7.3 is to show the high level of interconnectivity between the Quotas and Boundaries. Figure 7.4 depicts the direct relationship between the PBs

---

[2] The full list of critical pressures and how these were distilled into the Planetary Quota metrics is available in Meyer (2018).

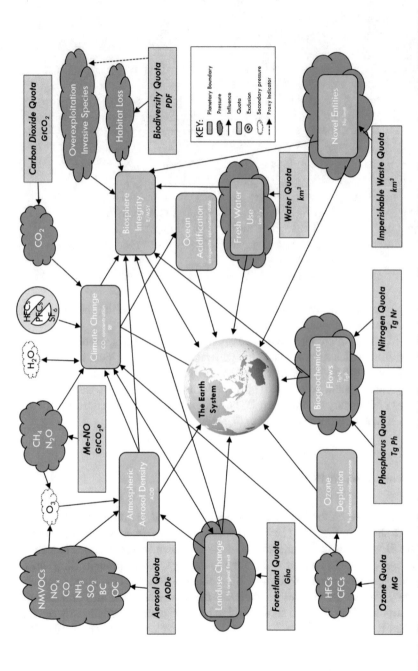

Fig. 7.3 The Planetary Boundaries define key processes which influence the Earth system. These are distilled into Pressures through the UN DPSIR framework. Critical pressures are shown in orange bubbles—grouped for equivalence with respect to the PBs. Excluded pressures are shown in a pale orange crossed-out circle. Secondary pressures are shown in white. PQs for each set of pressures are shown in green boxes. Causal relationships are shown with arrows

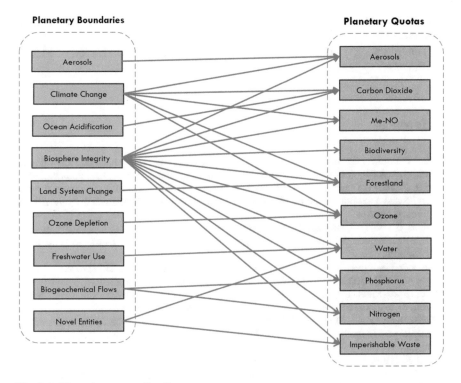

Fig. 7.4 PBs and corresponding Quotas

and each PQ. The Planetary Boundary authors identified climate change and biosphere integrity as "core Boundaries"—those with the highest interconnectivity with other Boundaries (Steffen et al. 2015). This interconnectivity is demonstrated in Fig. 7.4.

7.5   Conclusions

Research and development in multidisciplinary fields requires new research methodologies and different ways of thinking. Challenges such as limiting value judgements and managing system uncertainties were addressed in this project through integrated research and extended peer community engagement with international experts across the different disciplines reached in this research.

The DPSIR framework provides a mechanism through which to translate the Planetary Boundaries into pressure metrics that are more useful for policy, behaviour, and business applications. The framework has been used in the past to downscale the Boundaries to national level frameworks. However, the interconnectivity between the Boundaries has not previously been addressed. Our novel approach enabled the derivation of global but scalable limits for human pressures on the environment—the Planetary Quotas.

The following chapters describe the basis for and detailed derivation of each of the Planetary Quotas.

References

Brandt P, Ernst A, Gralla F, Luederitz C, Lang DJ, Newig J, Reinert F, Abson DJ, von Wehrden H (2013) A review of transdisciplinary research in sustainability science. Ecol Econ 92:1–15

Dao H, Peduzzi P, Chatenoux B, De Bono A, Schwarzer S, Friot D (2015) Environmental limits and Swiss footprints based on planetary boundaries. UNEP/GRID-Geneva & University of Geneva, Geneva

Douglas G (2005) A new approach to development- integrative improvement (II)™: sustainable development as if people and their physical, social and cultural environments mattered. EconWPA

Funtowicz SO, Ravetz JR (1993) Science for the post-normal age. Futures 25:739–755

Meyer K (2017) AirPhD. https://airphd.wordpress.com/. Accessed 2 Mar 2019

Meyer K (2018) Planetary Quotas and the Planetary Accounting Framework - comparing human activity to global environmental limits. PhD Thesis, Curtin University

Nykvist B, Persson Å, Moberg F, Persson LM, Cornell SE, Rockström J (2013) National environmental performance on planetary boundaries: a study for the Swedish environmental protection agency. Swedish environmental protection agency, Sweden

Ravetz JR (1999) What is post-normal science. Futures 31:647–653

Rockström J, Steffen W, Noone K, Persson A, Chapin FS III, Lambin E, Lenton TM, Scheffer M, Folke C, Schellnhuber HJ, Nykvist B, De Wit CA, Hughes T, van der Leeuw S, Rodhe H, Sorlin S, Snyder PK, Costanza R, Svedin U, Falkenmark M, Karlberg L, Corell RW, Fabry VJ, Hansen J, Walker B, Liverman D, Richardson K, Crutzen P, Foley J (2009) Planetary boundaries: exploring the safe operating space for humanity. Ecol Soc 14:32

Steffen W, Richardson K, Rockström J, Cornell SE, Fetzer I, Bennett EM, Biggs R, Carpenter SR, De Vries W, De Wit CA, Folke C, Gerten D, Heinke J, Mace GM, Persson LM, Ramanathan V, Reyers B, Sörlin S (2015) Planetary Boundaries: guiding human development on a changing planet. Science 347:736, 1259855

Stocker L, Burke GL (2017) Methods for sustainability research. Edward Elgar Publishing, Cheltenham, UK

Chapter 8
A Planetary Quota for Carbon Dioxide

> *We're running the most dangerous experiment in history right now, which is to see how much carbon dioxide the atmosphere… can handle before there is an environmental catastrophe.*
> Elon Musk

Abstract Carbon dioxide emissions are the primary human pressure on the environment causing the global average temperature to increase. There are many other human factors which also contribute to global temperature change in both positive (warming) and negative (cooling) ways. These include the emissions of other greenhouse gases, the changing reflectivity of Earth's surface due to land-use change, and the cooling effect of suspended particles in the atmosphere emitted during human activity.

There are two limits in the Planetary Boundary framework which address climate change. The first is for a maximum change in radiative forcing of ± 1 W/m^2, the energy balance in the atmosphere, to account for the many drivers of climate change. The second is for the concentration of carbon dioxide in the atmosphere of ≤ 350 ppm. The explicit limit for carbon dioxide in addition to the limit for radiative forcing highlights the importance of this gas to the function of the planet.

The current concentration of carbon dioxide in the atmosphere is ≥ 400 ppm, i.e. the limit has been exceeded. Thus, to return to the Planetary Boundary level as part of the "safe operating space", carbon dioxide will need to be withdrawn from the atmosphere.

There are several proposals in the academic literature for ways to reduce the concentration of carbon dioxide in the atmosphere to 350 ppm. The fastest pathway, and the only one that achieves 350 ppm within this century, entails aggressive reductions in carbon dioxide emissions of 15% per year from 2020, an average uptake of carbon dioxide into the landscape or some technological sequestration of 7.3 GtCO$_2$ each year from 2050 to 2080, and net zero emissions beyond 2080.

The Planetary Quota for carbon dioxide is thus *net carbon emissions* ≤ -7.3 GtCO$_2$/year. This limit can be compared to the net carbon footprint of any

© Springer Nature Singapore Pte Ltd. 2020
K. Meyer, P. Newman, *Planetary Accounting*,
https://doi.org/10.1007/978-981-15-1443-2_8

scale of human activity. If emission reductions in the scale of 15% per annum do not start by 2020, this limit will need to be revised to reach 350 ppm this century.

8.1   Introduction

Carbon dioxide (CO_2) emissions affect many of the Planetary Boundary thresholds, in particular the threshold for the atmospheric concentration of CO_2. CO_2 emissions are probably the single most important environmental impact humans are having on the planet.

This chapter begins with an introduction to the carbon cycle and a discussion about why there needs to be a specific Planetary Quota for CO_2 emissions, as opposed to one that combines all greenhouse gases. This is followed by an overview of ways to measure CO_2 emissions and the argument for selecting *net carbon footprint* as the indicator. The chapter goes on to discuss the Quota for carbon dioxide with respect to each of the corresponding PBs listed above. The chapter concludes with a discussion of current emissions with respect to the Quota and some examples of what might be needed to make the changes required to live within this limit.

8.2   Background

8.2.1   The Carbon Cycle

The Earth system has several biogeochemical cycles, in which a chemical substance moves between the biosphere (life on Earth), lithosphere (Earth's crust), atmosphere (the layer of gases surrounding the Earth), and hydrosphere (surface and atmospheric water). Carbon, oxygen, nitrogen, water, and phosphorus all have biogeochemical cycles—many of which are described in this book.

The carbon cycle (as shown in Fig. 8.1) can be viewed as two separate but linked cycles. In the first cycle, carbon moves relatively quickly between the atmosphere, ocean, ocean sediments, vegetation, soil, and freshwater. In the second cycle, carbon contained in dead plant and animal matter is buried under layers of sediment and over hundreds of thousands of years gradually turns to coal, oil, and natural gas (Ciais et al. 2013).

From the beginning of the Holocene epoch[1] until the Industrial Revolution approximately 11,500 years later, the concentration of CO_2 remained relatively constant at approximately 280 parts of CO_2 per million parts of atmosphere (ppm). Since the Industrial Revolution, human activity, in particular the burning of fossil

---

[1] The Holocene epoch is a period of time that started approximately 10,000 years ago. See Chap. 2 for more detail.

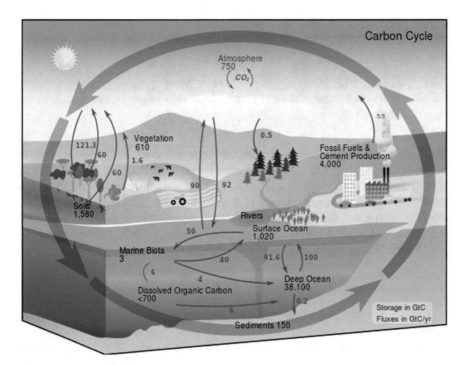

Fig. 8.1 The carbon cycle. (Saff 2008, PD (PD: This image has been released to the public domain))

fuels, the manufacture of cement, and deforestation, has led to a rapid increase in the concentration of CO_2 in the atmosphere. The atmospheric concentration of CO_2 is now over 400 ppm (NOAA 2018a). Concentrations have not been this high in at least 800,000 years (WMO 2017). Some estimates suggest that it has not been this high for 15 million years (Tripati et al. 2009). It is practically certain that concentrations of CO_2 have not been this high in human history—which spans approximately 380,000 years.

Since the Industrial Revolution, in the same period that there has been a substantial increase in the atmospheric concentration of CO_2, there has been an increase in global average temperature of just over 1 °C (NASA 2017). The concern over increasing levels of CO_2 in the atmosphere is that there is very strong evidence that the atmospheric concentration of CO_2 is linked to global average temperature (see Chap. 2, Sect. 2.3.1). There is a time lag between CO_2 emissions and temperature change so that even if we stopped emitting CO_2 today, the world would continue warming for some time before temperatures reached equilibrium again. The more total CO_2 emitted, the more warming we are likely to experience.

Historic data suggests that average temperatures and CO_2 concentrations do not change gradually and linearly. In the past, changes have started gradually until a tipping point has been reached at which point changes have become very rapid (see

Chap. 2, Sect. 2.3.1). If we reach a certain threshold of CO_2 concentration, we risk putting into motion dramatic and potentially irreversible change to the Earth system.

8.2.2   Why Separate Carbon Dioxide from Other Greenhouse Gases?

There are many human factors that contribute to climate change. Anthropogenic emission of carbon dioxide (CO_2) is only one factor, but it is also the greatest (IPCC 2013c) and needs particularly careful management (Steffen et al. 2015). Other human factors that influence climate change include greenhouse gas (GHG) emissions, stratospheric and tropospheric ozone, water vapour, surface albedo (the reflectivity of Earth's surface), and atmospheric aerosols (IPCC 2013b). There are two Planetary Boundary limits for climate change:

- Atmospheric concentration of CO_2 $\leq$350 ppm
- Radiative forcing $\leq\pm1.0$ W/m^2

The limits correspond to a temperature increase of approximately 1.7 °C above preindustrial times (Hansen et al. 2008). The first limit pertains solely to CO_2. The second provides a more all-encompassing limit that considers not only CO_2 but also the other factors which influence climate change. When the Planetary Boundaries were first published, the concentration of CO_2 in the atmosphere was 387 ppm

---

Box 8.1 Why 350 ppm?

The current concentration of CO_2 in the atmosphere is well over 400 ppm, yet we are not experiencing catastrophic change. This begs the question "Why is the PB for CO_2 concentration set at 350 ppm?"

There is a lag between the change in concentration of CO_2 and the change in global temperature. This means that even if we stopped emitting any greenhouse gases today, the world would continue warming for some time.

The PB for atmospheric concentration of CO_2 has been set at a level which is thought to minimise the risk of "highly non-linear, possibly abrupt and irreversible" change. This is based on data which suggests that the planet was mostly free of ice until CO_2 concentrations fell to somewhere between 350 and 550 ppm (Hansen et al. 2008).

The authors of the PBs justify locating the limit on the lower end of this range because there is evidence that the Earth's subsystems are already starting to behave differently than they did in a Holocene state (Rockström et al. 2009b). Moreover, the authors' review of existing climate models led them to believe that the models do not adequately take into account the severity of feedback loops (Rockström et al. 2009b).

(Rockström et al. 2009a). Now, in 2019, the concentration is over 413 ppm (Keeling et al. 2019). The change in radiative forcing since preindustrial times is approximately +1.5 W/m² (Myhre et al. 2013a) (Box 8.1).

Climate change is complex. Human activities contribute both warming and cooling effects also known as climate forcing effects. Greenhouse gas emissions warm the atmosphere by absorbing infrared radiation and thus trapping heat in the atmosphere. Atmospheric aerosols (tiny particles suspended in the atmosphere) can both warm the atmosphere by trapping heat and cool the atmosphere by reflecting it. Changing the make-up of Earth's surface can warm or cool the atmosphere by changing the albedo (or reflectivity). For example, the less surface area covered by ice (the most reflective surface on Earth), the more warming occurs. Even within a single "type" of climate change factor, for example, GHG emissions, there is not one single mechanism for warming. Carbon dioxide behaves differently to methane which behaves differently to nitrous oxide. The effects of different forcing elements are shown in Fig. 8.2.

To simplify the situation, two main indicators have been developed to allow the direct comparison of different forcing agents.

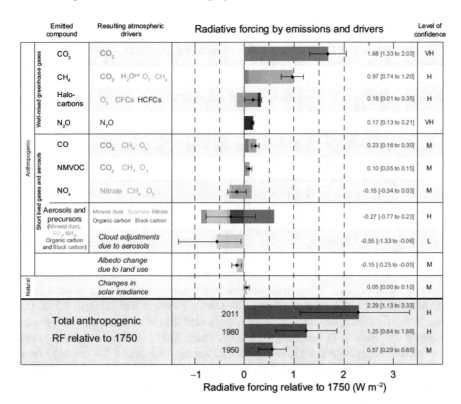

Fig. 8.2 Radiative forcing impacts of different compounds and atmospheric drivers resulting from the emitted compounds relative to 1750. (Figure SPM.5 from IPCC 2013c (IPCC allows the reuse of a small number of figures without formal permissions with appropriate acknowledgement))

1. Radiative forcing (RF): Radiative forcing is defined by the IPCC (2013c) as the change in energy flux at the top of the atmosphere caused by a forcing agent. Radiative forcing can be determined based on a change in concentration of a substance in the atmosphere or based on the amount of emissions of the substance (Pierrehumbert 2014). It can also be used to assess forcing effects due to the albedo of clouds and Earth's surface. It is measured in Watts per square metre. This metric is very useful as it allows the effects of greenhouse gases to be compared to the effects of aerosols and change in albedo as shown in Fig. 8.2. There is a proportional relationship between the change in equilibrium surface temperature and radiative forcing (Myhre et al. 2013b).

2. Global warming potential (GWP): This unit allows emissions of GHGs to be communicated in terms of *equivalent* kilograms (or tonnes) of CO_2 (CO_2e). The equivalency is determined by the global warming potential (GWP) of each gas. The GWP is the amount of heat trapped by a substance in a specified timeframe (typically 100 years), compared to the amount of heat trapped by the same amount of carbon dioxide. The GWP of nitrous oxide is 256–298 over a timescale of 100 years. This means that burning 1 kg of nitrous oxide is expected to have a similar warming effect over 100 years as burning between 256 and 298 kg of carbon dioxide.

It is a common practice to use GWP to measure greenhouse gases collectively (Greenhalgh et al. 2005). After the Kyoto Protocol was initiated, the practice of GHG accounting—measuring and reporting greenhouse gases—became formalized and standardized so that countries could demonstrate whether or not they were meeting their commitments under the protocol and be held accountable for this. In formal GHG accounting procedures, the impacts of carbon dioxide (CO_2), methane (CH_4), nitrous oxide (N_2O), hydrofluorocarbons (HFCs), perfluorocarbons (PFCs), sulphur hexafluoride (SF_6), and nitrogen trifluoride (NF_3) must all be accounted for.

Although GWP and radiative forcing imply a certain equivalency between different forcing agents and particularly between different greenhouse gases, the impacts are not equivalent. As shown in Fig. 8.2, CO_2 has had the greatest contribution to climate forcing since 1750. It will also continue to have the greatest contribution, not only because of the high level of past and predicted emissions but also because we will continue to experience the warming effects of past CO_2 emissions for tens of thousands of years. As discussed above, the carbon cycle has a rapid sub-cycle and a slow sub-cycle. The problem is that in the last 50–100 years, humans have extracted billions of tonnes of carbon from the slow cycle and released it into the fast cycle, disturbing the natural balance. To restore the balance, the carbon we have released will need to be returned to the slow cycle. This will happen naturally over tens of thousands of years. Human interventions can accelerate CO_2 removal from the atmosphere through revegetation and reforestation (IPCC 2013c), or through geoengineering (Druckman and Jackson 2010), though not at the rates we are currently emitting it.

Carbon dioxide, nitrous oxide, methane, and some halocarbons are all considered to be "long-lived" greenhouse gases. However, the "long" in "long-lived" varies substantially. The atmospheric lifetime of methane is estimated to be 7–11 years (IPCC 2013a) and nitrous oxide, 118–131 years (Volk et al. 1997; Hsu and Prather 2010; Fleming et al. 2011). The warming potential of these gases is higher per kilogram that CO_2. However, emitting these substances does not have the same long-term commitment to warming as emitting CO_2. It is thus not equivalent to emit 1 kg of nitrous oxide to 298 kg of CO_2 even though the immediate warming effects are comparable. To have a low risk of departure from a Holocene-like state, and to respect the Planetary Boundaries, which have a specific CO_2 concentration limit, CO_2 will need to be removed from the atmosphere. Thus, a specific Planetary Quota for CO_2 is required.

8.3   An Indicator for Carbon Dioxide Emissions

The term carbon footprint originated from the concept of an Ecological Footprint (Sundha and Melkania 2016). In its broadest sense, a carbon footprint is the amount of carbon (or an equivalent amount of greenhouse gas emissions) emitted by a group or over the course of an activity (e.g. the manufacturing of a product). Carbon footprint tools are extensively used in a wide range of applications from basic online calculators to detailed life-cycle analyses of products, regions or nations (Sundha and Melkania 2016; Wright et al. 2011). The concept of a carbon footprint forms the underpinning of carbon accounting. Specific carbon footprint definitions range from a measurement of CO_2 emissions to a measurement of all greenhouse gas emissions and can be reported in terms of tonnes of CO_2 and tonnes of CO_2 equivalent or sometimes in terms of the forest area which would be required to absorb the CO_2 (Cucek et al. 2012). It is a common practice to report gross emissions of CO_2 and uptake of CO_2 from land-use change and forestry separately. However, combing these figures is sometimes done to report *net carbon dioxide emissions*. Given that a negative budget will be needed to return to 350 ppm, we have chosen *net carbon dioxide emissions* as the indicator for the Planetary Quota for carbon dioxide.

8.4   The Limit

CO_2 emissions do not only affect the two Planetary Boundary metrics for climate change (concentration of CO_2 and radiative forcing). CO_2 emissions also influence the Planetary Boundaries for ocean acidification, biosphere integrity, and land-use change. As such, there are five Planetary Boundary limits that were considered when deriving the PQ for CO_2:

- Concentration of CO_2 in the atmosphere ≤ 350 ppm
- Radiative forcing ≤ 1 W/m^2
- Saturation state of aragonite ≤ 2.75

- Extinction rate $\leq$10 extinctions per million species per year
- Remaining forestland $\geq$75% original forest area

Each of these is discussed below.

8.4.1   Total Radiative Forcing

As discussed previously, total radiative forcing is the most holistic measure of global warming/cooling and accounts for all GHG emissions as well as change in albedo (reflectivity of Earth's surface which changes with land-use change) and the emission of aerosols (particles suspended in the atmosphere). Emissions of CO_2 alone have resulted in a change of RF of 1.68 W/m^2 since the Industrial Revolution (IPCC 2013c). There are four Planetary Quotas which will influence total radiative forcing: the PQs for carbon dioxide, methane and nitrous oxide ("Ag-GHGs"), reforestation, and aerosols. It is thus not possible to derive a specific limit for any one factor which contributes to radiative forcing in isolation. The effects on radiative forcing of all four limits were considered collectively for the original Quotas in Meyer and Newman (2018) and Meyer (2018). It is assumed that the amendments to the Quotas for CO_2 and Ag-GHGs presented in this book have a negligible effect on their combined radiative forcing in the long term.

8.4.2   Ocean Acidification

The oceans absorb CO_2 from the atmosphere at a rate that is loosely proportional to the concentration of CO_2 in the atmosphere. This process has significantly dampened the warming effects of anthropogenic CO_2 emissions. However, the absorption of CO_2 by the oceans has also increased the ocean pH levels. This means that the seas are more acidic which has disastrous impacts on marine ecosystems. Marine ecosystems are critical to the functioning of the Earth system (Rockström et al. 2009a), and thus ocean acidity is an important Planetary Boundary. However, provided the PB for CO_2 concentration is respected, the PB for ocean acidification will also be respected (Steffen et al. 2015). As such, it follows that provided PQ for CO_2 respects the PB for CO_2 concentration, the PB for ocean acidification will be intrinsically respected.

8.4.3   Biosphere Integrity

Climate change is one of the five key pressures leading to the loss of species (Secretariat of the CBD 2001). There is no specific concentration of CO_2 or level of climate change considered "safe" with respect to extinction rate. It is assumed that the PB limits for climate change are adequate to address species loss due to climate change.

8.4.4 CO_2 Concentration

The amount of CO_2 in the atmosphere is the single greatest driver of climate change. However, the more popular metric for climate change is global average temperature change. For many years, an upper limit of 2 °C was considered the maximum "safe" amount of warming. However, in more recent years, as our understanding of climate science has improved, a more stringent upper limit of 1.5 °C has been more broadly accepted. While there is not an exact relationship between the Planetary Boundary for atmospheric CO_2 concentration of ≤ 350 ppm and the warming limit of 1.5 °C, they describe roughly the same target.

In past publications introducing Planetary Accounting and the Planetary Quotas (Meyer and Newman 2018; Meyer 2018), a rapid decarbonization pathway proposed by Hansen et al. (2013) was used as the basis for the Planetary Quota for CO_2 emissions. The proposed Quota was -7.3 $GtCO_2$/year by 2030. Hansen's pathway was then and is still today the most ambitious decarbonization pathway in the literature and the only one which explicitly returns CO_2 concentrations to 350 ppm this century.

On October 2018 the Intergovernmental Panel for Climate Change released a special report on global warming of 1.5 °C that includes a review of 1.5 °C pathways (IPCC 2018a). The authors of this report define 1.5 °C pathways as pathways that give a one in two or better chance of limiting warning to 1.5 °C or returning to stabilized temperatures of 1.5 °C by 2100 (Allen et al. 2018). They warn that the predicted impacts associated with exceeding and then returning to 1.5 °C are far greater than those predicted for pathways where temperatures do not exceed 1.5 °C (Allen et al. 2018). The pathways reviewed in this report do not explicitly define the date at which CO_2 concentrations would return to 350 ppm. However, the basis of the 350 ppm target and the 1.5 °C target is the same (i.e. to limit the risk of non-linear, runaway climate change) (Rockström et al. 2009a; IPCC 2018a).

There were 90 scenarios analysed in the IPCC report under the pathway group "1.5 °C or 1.5 °C consistent". Of these, 9 had a 50–66% likelihood of limiting warming to below 1.5 °C during the twenty-first century. No pathways have a greater than 66% likelihood of this outcome (Rogelj et al. 2018).

For a 66% probability of remaining below 1.5 °C, and accounting for the risk of carbon release from future permafrost thawing and methane release from wetlands, the remaining carbon budget from the end of 2017 is 320 $GtCO_2$ (Rogelj et al. 2018).

The IPCC special report does not propose a specific pathway with which to allocate this remaining budget over time. However, the author's acknowledge that the reductions required to limit warming to 1.5 °C depend heavily on the year in which the reductions begin (Rogelj et al. 2018). It has been estimated that for every year's delay in initiating reductions, the time available to achieve zero emissions reduces by 2 years (Allen and Stocker 2013; Leach et al. 2018).

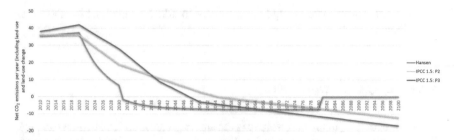

Fig. 8.3 Comparison of Hansen et al. (2013) and two IPCC (2018a) 1.5 °C decarbonization pathways

Four specific scenarios are provided in the IPCC 1.5 °C report to show the variation in mechanisms with which the budget could be achieved. All four pathways rely on a combination of rapid CO_2 emission reductions and carbon dioxide removal through either or both of Agriculture, Forestry, and Other Land Use (AFOLU) and Bioenergy with Carbon Capture and Storage (BECCS). The pathways include:

- P1/LED: A low energy demand scenario in which the carbon budget could be achieved without carbon uptake beyond afforestation.
- P2/S1: A sustainability-oriented scenario with low population growth, low consumption, high energy efficiency, and minimal BECCS.
- P3/S2: A middle of the road scenario where societal and technological development follows historic patterns and reductions are predominantly achieved by changing the production of energy and products and substantial BECCS.
- P4/S5: A fossil fuel intensive and high energy demand scenario which requires very high uptake of CO_2 through BECCS.

Of these scenarios, P1, P2, and P3 are expected to have limited or no overshoot of 1.5 °C, while P4 has some overshoot. Figure 8.3 shows P2 and P3 against the Hansen scenario that was used as the basis for the original Quota for CO_2.[2]

8.4.4.1   Comparison of Pathways

All three pathways shown above fall within the total carbon budget for 2018 onwards of 320 $GtCO_2$. This means that any of these pathways give a two thirds or better chance of remaining below 1.5 °C. Cumulative emissions from 2018 to 2100, the year net zero emissions are achieved, and the maximum net CO_2 uptake for each of the scenarios are shown in (Table 8.1):

- Hansen: −50 $GtCO_2$
- IPCC 1.5 °C—P2: 315 $GtCO_2$

---

[2]Annual CO_2 emission data was not available for P1. P4 is excluded because it is expected to lead to overshoot of 1.5 °C.

Table 8.1 Comparing 350 ppm and 1.5 °C consistent pathways

	Net cumulative emissions (GtCO$_2$)	Year that net zero emissions are achieved	Maximum net CO$_2$ uptake (GtCO$_2$)
Hansen	−50	2030	7.3
IPCC 1.5 °C—P2	315	2055	12
IPCC 1.5 °C—P3	160	2048	17

- IPCC 1.5 °C—P3: 160 GtCO$_2$

The timeframe for the IPCC pathways to return CO$_2$ concentrations to 350 ppm is not specified; however, it would not be within this century given that the net cumulative emissions for these scenarios are positive.

Hansen's pathway requires steep emission reductions of 15% per annum from 2020 to 2050, with net uptake of 7.3 GtCO$_2$/year from 2050 to 2080, and net zero emissions beyond 2080. This pathway includes 100 GtC carbon uptake this century. In contrast, P2 and P3 require much less aggressive immediate emission reductions (5% and 3% per annum, respectively) but have a much heavier reliance on carbon uptake (257 GtCO$_2$ and 490 GtCO$_2$, respectively). The pathway that provides the least risk of non-linear runaway climate change is Hansen's. However, this is also the most challenging pathway and the one which provides the least space to meet social needs in poorer countries. P3 is the least challenging in the immediate term. However, it is also the riskiest pathway of the three, as it places a heavy reliance on BECCS technologies.

As such, we have revised the Quota for CO$_2$ based on the IPCC 1.5 °C P2 pathway to be *net CO$_2$ emissions ≤ -12 GtCO$_2$/year by 2100, with a maximum cumulative budget from 2018 to 2100 of 320 GtCO$_2$.*

8.5 Discussion

The PQ for CO$_2$ emissions of −12 GtCO$_2$/year is an ambitious target. Global annual emissions are currently in the order of 36 GtCO$_2$/year (B. Jackson et al. 2018). At a projected population of nine billion, the PQ for CO$_2$ equates to an average carbon footprint of −1.3 tCO$_2$/person/year. To put this into context, the average carbon footprint of the OECD countries is currently 11 tCO$_2$/person/year; an average car emits approximately 4.7 tCO$_2$/year; and a return flight from NZ to London for one person equates to emissions of approximately 7.5 tCO$_2$.

For 3 years in a row, from 2014 to 2016, the global CO$_2$ emissions remained relatively constant. In 2017 and 2018, emissions rose again (NOAA 2018b) mostly due to increased land clearing, meaning that CO$_2$ emissions have not yet peaked. However, the trend in the past 10 years suggests that the rate of increase in carbon emissions is dropping, especially in the use of fossil fuels.

Table 8.2 Examples of different scales of activity which have or could contribute to achieving the PQ for CO_2 emissions

Achieving the Planetary Quota for CO_2 emissions			
	Community	**Government**	**Business**
Large	Development of an independent global community to report on the scientific understanding of CO_2 emissions and corresponding impacts, *e.g. the Intergovernmental Panel for Climate Change*	Develop a global treaty to limit carbon emissions, *e.g. the Paris Agreement*	Develop innovative solutions for the removal of CO_2 from the atmosphere as in many large businesses and their R&D programmes
Medium	Lobby government to go carbon neutral as in many cities involved in C40, 100 Resilient Cities, and ICLEI	Implement rail, walk-ways, and cycle paths, commit to carbon neutral	Install solar panels to power business operations, go carbon neutral
Small	Install a community renewable energy system with battery storage, *e.g. White Gum Valley* (Wiktorowicz et al. 2018)	Create demonstrations of zero carbon buildings, put solar on all government buildings, electrify fleet	Go fossil fuel-free through roof top solar and electric vehicle fleets
Individual	Behaviour change: Choose a fossil fuel-free electricity provider Switch to an electric vehicle Turn off lights Eat less meat	Develop new models for a low-carbon economy, *e.g. a new financial model to make city rail affordable through land development* (Newman et al. 2018)	Disruptive low-carbon innovations, *e.g. multiple entrepreneurs have created new battery technology which is essential to accelerate the transition to a fossil fuel-free economy* (Newman 2018)

For further examples, see Newman et al. (2017)

Table 8.2 shows examples of how this PQ could be put into practice at different scales of activity and across different sectors. The examples relate to varying timeframes.

The timeframe to reduce net emissions to zero to meet the criteria for this PQ is only 10 years. This is very ambitious. However, the journey is not only starting now. People have been trying to manage CO_2 emissions for several decades already. Moreover, humanity has made dramatic changes in both technology and behaviour over very short periods of time before. For example, consider the mobile phone. The first portable phone—the DynaTAC—was available in the early 1980s. For 27 years, there were regular advancements—phones became smaller, batteries lasted longer. There was continual improvement at a fairly constant rate. Then in 2007 the first smart phone was introduced—the iPhone—an all-in-one phone, camera, music player, and Internet-enabled PDA. Now, approximately 10 years later, almost 90% of the world's population is covered by 2G networks, and over two billion people, more than a quarter of the world's population, own smart phones. This kind of disruptive innovation is now being seen with solar energy (especially rooftop solar) and wind power which between 2010 and 2016 increased by 41% and 16%, respectively (IPCC 2018b). Such changes will need to continue to exponentially increase in all sectors of the economy. For the removal of carbon from the atmosphere, a range of new technologies will be needed, as well as traditional techniques of carbon sequestration including forestry, agriculture, coastal management, and urban management (Newman et al. 2017).

8.6 Conclusion

The impacts of carbon dioxide emissions cannot be offset by reductions in other greenhouse gas emissions because of the long atmospheric lifetime of carbon dioxide. This is reflected in the Planetary Boundaries through the specific limit for the concentration of CO_2 in the atmosphere. The PB for CO_2 has been exceeded, as such the corresponding PQ is negative—i.e. carbon dioxide must be withdrawn from the atmosphere. The PQ for CO_2 is *net carbon emissions* ≤ -12 $GtCO_2$/year with a *net cumulative carbon budget from 2018 to 2100 of 320 GtCO$_2$*. This is based on a pathway that is estimated to have a 66% or better chance of keeping the global average surface temperature at or below 1.5 °C.

References

Allen MR, Stocker TF (2013) Impact of delay in reducing carbon dioxide emissions. Nat Clim Chang 4:23

Allen MR, Dube OP, Solecki W, Aragón-Durand F, Cramer W, Humphreys S, Kainuma M, Kala J, Mahowald N, Mulugetta Y, Perez R, Wairiu M, Zickfeld K (2018) Framing and context. In: Masson-Delmotte V, Zhai P, Pörtner HO, Roberts D, Skea J, Shukla PR, Pirani A, Moufouma-Okia W, Péan C, Pidcock R, Connors S, Matthews JBR, Chen Y, Zhou X, Gomis MI, Lonnoy E, Maycock T, Tignor M, Waterfield T (eds) Global warming of 1.5 °C. An IPCC special report on the impacts of global warming of 1.5 °C above pre-industrial levels and related global greenhouse gas emission pathways, in the context of strengthening the global response to the threat of climate change, sustainable development, and efforts to eradicate poverty. IPCC, Cambridge, UK

Ciais P, Sabine C, Bala G, Bopp L, Brovkin V, Canadell J, Chhabra A, Defries R, Galloway J, Heimann M, Jones C, Le Quéré C, Myneni RB, Piao S, Thornton P (2013) Carbon and other biogeochemical cycles. In: Stocker TF, Qin D, Plattner G-K, Tignor M, Allen SK, Boschung J, Nauels A, Xia Y, Bex V, Midgley PM (eds) Climate change 2013: the physical science basis. Contribution of working group I to the fifth assessment report of the Intergovernmental Panel on Climate Change. Cambridge University Press, Cambridge, UK

Cucek L, Klemes JJ, Kravanja Z (2012) A Review of Footprint analysis tools for monitoring impacts on sustainability. J Clean Prod 34:9–20

Druckman A, Jackson T (2010) The bare necessities: how much household carbon do we really need? Ecol Econ 69:1794–1804

Fleming E, Jackman C, Stolarski R, Douglass A (2011) A model study of the impact of source gas changes on the stratosphere for 1850–2100. Atmos Chem Phys 11:8515

Greenhalgh S, Broekhoff D, Daviet F, Ranganathan J, Acharya M, Corbier L, Oren K, Sundin H (2005) The GHG protocol for project accounting. Greenhouse gas protocol. World Resources Institute and World Business Council for Sustainable Development, Wahington, DC

Hansen J, Sato M, Kharecha P, Beerling D, Berner R, Masson-Delmotte V, Pagani M, Raymo M, Royer DL, Zachos JC (2008) Target atmospheric CO_2: where should humanity aim? Open Atmos Sci J 2:217–231

Hansen J, Kharecha P, Sato M, Masson-Delmotte V, Ackerman F, Beerling DJ, Hearty PJ, Hoegh-Guldberg O, Hsu S-L, Parmesan C, Rockstrom J, Rohling EJ, Sachs J, Smith P, Steffen K, van Susteren L, von Schuckmann K, Zachos JC (2013) Assessing dangerous climate change: required reduction of carbon emissions to protect young people, future generations and nature. PLoS One 8:e81648

Hsu J, Prather MJ (2010) Global long-lived chemical modes excited in a 3-D chemistry transport model: stratospheric N_2O, NO_y, O_3 and CH_4 chemistry. Geophys Res Lett 37

IPCC (2013a) Anthropogenic and natural radiative forcing. In: Stocker TF, Qin D, Plattner G-K, Tignor M, Allen SK, Boschung J, Nauels A, Xia Y, Bex V, Midgley PM (eds) Climate change 2013: the physical science basis. Contribution of the working group I to the fifth assessment report of the Intergovernmental Panel on Climate Change. IPCC, Cambridge, UK

IPCC (2013b) Climate change 2013: the physical science basis. Contribution of working group I to the fifth assessment report of the Intergovernmental Panel on Climate Change. Cambridge University Press, Cambridge, UK

IPCC (2013c) Summary for policymakers. In: Stocker TF, Qin D, Plattner G-K, Tignor M, Allen SK, Boschung J, Nauels A, Xia Y, Bex V, Midgley P (eds) Climate change 2013: the physical science basis. Contribution of working group I to the fifth assessment report of the Intergovernmental Panel on Climate Change. Cambridge University Press, Cambridge, UK

IPCC (2018a) Global warming of 1.5 °C. An IPCC special report on the impacts of global warming of 1.5 °C above pre-industrial levels and related global greenhouse gas emission pathways, in the context of strengthening the global response to the threat of climate change, sustainable development, and efforts to eradicate poverty. IPCC, Cambridge, UK

IPCC (2018b) Summary for policymakers. In: Masson-Delmotte V, Zhai P, Pörtner HO, Roberts D, Skea J, Shukla PR, Pirani A, Moufouma-Okia W, Péan C, Pidcock R, Connors S, Matthews JBR, Chen Y, Zhou X, Gomis MI, Lonnoy E, Maycock T, Tignor M, Waterfield T (eds) Global warming of 1.5 °C. An IPCC special report on the impacts of global warming of 1.5 °C above pre-industrial levels and related global greenhouse gas emission pathways, in the context of strengthening the global response to the threat of climate change, sustainable development, and efforts to eradicate poverty. World Meteorological Organisation, Geneva, Switzerland

Jackson RB, Le Quéré C, Andrew RM, Canadell JG, Korsbakken JI, Liu Z, Peters GP, Zheng B (2018) Global energy growth is outpacing decarbonization. International Project Office, Canberra, Australia

Keeling RF, Walker SJ, Piper SC, Bollenbacher AF (2019) Scripps CO2 Program. University of California, La Jolla, CA. http://scrippsco2.ucsd.edu

Leach NJ, Millar RJ, Haustein K, Jenkins S, Graham E, Allen MR (2018) Current level and rate of warming determine emissions budgets with ambitious mitigation. Nat Geosci 11:574–579

Meyer K (2018) Planetary Quotas and the Planetary Accounting Framework - comparing human activity to global environmental limits. PhD Thesis, Curtin University

Meyer K, Newman P (2018) The Planetary Accounting Framework: a novel, quota-based approach to understanding the planetary impacts of any scale of human activity in the context of the Planetary Boundaries. Sustain Earth 1:4

Myhre G, Shindell D, Bréon F-M, Collins W, Fuglestvedt J, Huang J, Koch D, Lamarque J-F, Lee D, Mendoza B, Nakajima T, Robock A, Stephens G, Takemura T, Zhang H (2013a) Anthropogenic and natural radiative forcing. In: Stocker TF, Qin D, Plattner G-K, Tignor M, Allen SK, Boschung J, Nauels A, Xia Y, Bex V, Midgley PM (eds) Climate change 2013: the physical science basis. Contribution of working group I to the fifth assessment report of the Intergovernmental Panel on Climate Change. Cambridge University Press, Cambridge, UK

Myhre G, Shindell D, Breon F-M, Collins W, Fuglestvedt J, Huang J, Koch D, Lamarque JL, Lee B, Mendoza B, Nakajima T, Robock A, Stephens G, Takemura T, Zhan H (2013b) Anthropogenic and natural radiative forcing. In: Stocker TF, Qin D, Plattner G-K, Tignor M, Allen SK, Boschung J, Nauels A, Xia Y, Bex V, Midgley PM (eds) Climate change 2013: the physical science basis. Contribution of working group I to the fifth assessment report of the Intergovernmental Panel on Climate Change. Cambridge University Press, Cambridge, UK

NASA (2017) NASA, NOAA data show 2016 warmest year on record globally. Washington, DC, NASA

Newman P (2018) The renewable city revolution. In: Droege P (ed) Urban energy transitions, 2nd edn. Elsevier, Berlin

Newman P, Beatley T, Boyer H (2017) Resilient cities: overcoming fossil fuel dependence. Island Press/Center for Resource Economics, Washington, DC

Newman P, Davies-Slate S, Jones E (2018) The Entrepreneur Rail Model: funding urban rail through majority private investment in urban regeneration. Res Transp Econ 67:19–28

NOAA (2018a) Trends in atmospheric carbon dioxide. https://www.esrl.noaa.gov/gmd/ccgg/trends/monthly.html. Accessed 15 May 2018

NOAA (2018b) Trends in atmospheric carbon dioxide. National Oceanic & Atmospheric Administration - Earth System Research Laboratory - Global Monitoring Division. https://www.esrl.noaa.gov/gmd/ccgg/trends/full.html. Accessed 10 Jan 2018

Pierrehumbert RT (2014) Short-lived climate pollution. Annu Rev Earth Planet Sci 42:341–379

Rockström J, Steffen W, Noone K, Persson A, Chapin FS, Lambin E, Lenton TM, Scheffer M, Folke C, Schellnhuber HJ, Nykvist B, De Wit CA, Hughes T, van der Leeuw S, Rodhe H, Sorlin S, Snyder PK, Costanza R, Svedin U, Falkenmark M, Karlberg L, Corell RW, Fabry VJ, Hansen J, Walker B, Liverman D, Richardson K, Crutzen P, Foley J (2009a) Planetary Boundaries: exploring the safe operating space for humanity. Ecol Soc 14:32

Rockström J, Steffen W, Noone K, Persson A, Chapin FS, Lambin E, Lenton TM, Scheffer M, Folke C, Schellnhuber HJ, Nykvist B, De Wit CA, Hughes T, van der Leeuw S, Rodhe H, Sorlin S, Snyder PK, Costanza R, Svedin U, Falkenmark M, Karlberg L, Corell RW, Fabry VJ, Hansen J, Walker B, Liverman D, Richardson K, Crutzen P, Foley J (2009b) Planetary Boundaries: exploring the safe operating space for humanity - supplementary information. Ecol Soc 14:32

Rogelj J, Shindell D, Jiang K, Fifita S, Forster P, Ginzburg V, Handa C, Kheshgi H, Kobayashi S, Kriegler E, Mundaca L, Séférian R, Vilariño MV (2018) Mitigation pathways compatible with 1.5 °C in the context of sustainable development. In: Masson-Delmotte V, Zhai P, Pörtner HO, Roberts D, Skea J, Shukla PR, Pirani A, Moufouma-Okia W, Péan C, Pidcock R, Connors S, Matthews JBR, Chen Y, Zhou X, Gomis MI, Lonnoy E, Maycock T, Tignor M, Waterfield T (eds) Global warming of 1.5 °C. An IPCC special report on the impacts of global warming of 1.5 °C above pre-industrial levels and related global greenhouse gas emission pathways, in the context of strengthening the global response to the threat of climate change, sustainable development, and efforts to eradicate poverty. IPCC, Cambridge, UK

Saff K (2008) Carbon cycle-cute diagram.svg. Wikimedia

Secretariat of the CBD (2001) Global biodiversity outlook 1. UNEP, CBD, Montreal

Steffen W, Richardson K, Rockström J, Cornell SE, Fetzer I, Bennett EM, Biggs R, Carpenter SR, De Vries W, De Wit CA, Folke C, Gerten D, Heinke J, Mace GM, Persson LM, Ramanathan V, Reyers B, Sörlin S (2015) Planetary Boundaries: guiding human development on a changing planet. Science 347:736, 1259855

Sundha P, Melkania U (2016) Carbon footprinting: a tool for environmental management. Int J Agric Environ Biotechnol 9:247–257

Tripati AK, Roberts CD, Eagle RA (2009) Coupling of CO_2 and ice sheet stability over major climate transitions of the last 20 million years. Science 326:1394–1397

Volk C, Elkins J, Fahey D, Dutton G, Gilligan J, Loewenstein M, Podolske J, Chan K, Gunson M (1997) Evaluation of source gas lifetimes from stratospheric observations. J Geophys Res 102:25543–25564

Wiktorowicz J, Babaeff T, Breadsell J, Byrne J, Eggleston J, Newman P (2018) WGV: an australian urban precinct case study to demonstrate the 1.5 °C agenda including multiple SDGs.
Urban Plan 3:64–81

WMO (2017) WMO greenhouse gas bulletin. WMO, Geneva

Wright LA, Kemp S, Williams I (2011) 'Carbon footprinting': towards a universally accepted
definition. Carbon Manag 2:61–72

Chapter 9
A Quota for Agricultural GHG Emissions (Methane and Nitrous Oxide)

> There's one issue that will define the contours of this century
> more dramatically than any other, and that is the urgent threat
> of a changing climate
> Barack Obama

Abstract Excessive human emissions of greenhouse gases (GHGs) are one of the "forcing" factors that has contributed to a change in global average temperatures by changing the balance of radiation at the top of Earth's atmosphere. Other anthropogenic forcing factors include change in albedo (Earth's reflectivity) due to land-use change and the emissions of aerosols and aerosol precursors. There is a Planetary Boundary for the change in radiative forcing since preindustrial times of $\leq \pm 1$ W/m^2.

Carbon dioxide, methane, nitrous oxide, and halocarbons are called "long-lived" or "well-mixed" gases. This means that they remain in the atmosphere long enough that the location of the source of the emissions is irrelevant. The impacts are experienced on a global scale. The warming effects of long-lived gases can all be expressed in terms of equivalent emissions of carbon dioxide (CO$_2$e). This chapter is about methane and nitrous oxide. The basis of the limit proposed for these gases is that remaining emissions would come almost entirely from agriculture and hence we have called them agricultural GHG emissions or Ag-GHGs.

To operate within the Planetary Boundaries will require a net withdrawal of carbon dioxide from the atmosphere and a phase out of halocarbons to zero emissions. In contrast, it is possible to continue to emit a small amount of methane and nitrous oxide without exceeding the Planetary Boundaries. Moreover, based on current prediction, it is anticipated that small amounts of these emissions will remain necessary in order to feed the global population. If carbon dioxide, halocarbons, methane, and nitrous oxide emissions were considered collectively, it would be conceivable that the Planetary Quota could be met without withdrawal of carbon dioxide or a complete phase out of halocarbons. As such, these pressures are addressed separately, and only methane and nitrous oxide are considered collectively.

The limit for methane and nitrous oxide emissions is based on the 2100 targets for these gases under the most stringent emission reduction pathway proposed by the Intergovernmental Panel on Climate Change (IPCC) in their fifth Assessment Report and was reinforced in the IPCC Special Report on 1.5 °C. The IPCC figures are based on almost eliminating nonagricultural emissions of both gases and optimizing the conflicting goals of minimizing emissions while maximizing agricultural output per land area in order to meet global food demands. As such, we have collectively termed these emissions agricultural GHGs, or Ag-GHGs.

The PQ for Ag-GHGs is *net emissions* ≤ 5 $GtCO_2e$/year. This limit can be compared to the net emissions of Ag-GHGs associated with any scale of human activity—the "Ag-GHG footprint".

9.1 Introduction

Methane and nitrous oxide are two of the four "well-mixed" greenhouse gases (GHGs) that are considered within the Planetary Quotas. The other two are carbon dioxide and halocarbons. Carbon dioxide emissions are addressed through a specific Quota because of their long atmospheric lifetime and because there is a specific Planetary Boundary for the concentration of carbon dioxide in the atmosphere (see Chap. 8). Moreover, because the limit for atmospheric carbon dioxide has already been exceeded, this Planetary Quota must be negative. To meet the Planetary Boundary for ozone depletion, halocarbons must be phased out entirely. In contrast, it is possible to return to the Planetary Boundaries corresponding to methane and nitrous oxide emissions without completely phasing these pressures out. Thus, while there are indicators that enable these pressures to be considered collectively, doing so would not equate to a true translation of the Planetary Boundaries to Planetary Quotas. Thus, only methane and nitrous oxide are considered together. A major anthropogenic source of both methane and nitrous oxide, and the basis for a greater than zero Quota, is agriculture. We thus refer to these gases collectively as "agricultural GHGs" (Ag-GHGs) from here on.

The emissions of Ag-GHGs are two of the biggest contributors to global climate change (IPCC 2014). They are emitted during the combustion of fossil fuels, from agriculture and agricultural practices, and because of some types of land use and land-use changes (Ciais et al. 2013). Current concentrations of these gases exceed levels measured for at least 800,000 years, and the rate of change of emissions increased more in the last 100 years than any rate over the past 20,000 years (Ciais et al. 2013).

This chapter begins with an introduction to the methane and nitrous oxide cycles and a brief history of human use of these gases. This is followed by the rationale behind the indicator used for this Quota and the scientific basis for the PQ limit. The chapter concludes with a discussion about the PQ for Ag-GHGs and the types of actions that may help humanity to live within this.

9.2   The Methane Cycle

Methane (CH_4) is generally considered to be the second most important greenhouse gas after carbon dioxide. The warming potential of nitrous oxide is higher than that of methane per kilogram of emission. However, human activities are responsible for a substantially higher amount of methane emissions. The radiative forcing of methane since preindustrial times is approximately 0.97 W/m^2 (Myhre et al. 2013). This means that since 1750, methane emissions have increased the energy flux at the top of Earth's atmosphere by approximately 1 W/m^2. To put this into context, the total change in radiative forcing since 1750 is approximately +1.5 W/m^2 (Myhre et al. 2013). The change in radiative forcing due to carbon dioxide is 1.68 W/m^2 and from nitrous oxide emissions it is 0.18 W/m^2.

Methane is a naturally occurring gas in Earth's atmosphere. Before the Industrial Revolution, methane made up approximately 722 parts per billion of atmosphere (ppb). Biological methane is emitted during the fermentation of organic matter in low oxygen conditions. Natural biological methane is emitted by wetlands, bacteria, termites, and a small amount from the oceans. There are also natural sources of fossil methane (methane stored under Earth's crust). This can be released into the atmosphere via terrestrial leaks, geothermal vents, and volcanic eruptions.

Methane is removed from the atmosphere naturally through photochemistry with hydroxyl radicals. It is thought that smaller amounts of methane are also removed through reactions with chlorine and oxygen radicals, by oxidation in aerated solids, and through reactions with chlorine in the marine boundary layer (Allan et al. 2007).

The greatest anthropogenic source of methane is biological emissions from agriculture and land-use change. Rice paddies, livestock, landfill off-gassing, man-made lakes and wetlands, and waste treatment plants are all examples of biological anthropogenic sources of methane. Fossil methane can be leaked during the extraction and use of fossil fuels. Anthropogenic fires, i.e. burning plant biomass, also releases methane emissions (Conrad 1996).

Between 1750 and 2011, methane concentrations have increased from 722 to 1803 ppb (Myhre et al. 2013). From 1750 until the mid-1980s, the increase was almost exponential. However, from the mid-1980s, the atmospheric concentrations of methane stayed relatively constant for about 20 years. This is thought to have been caused by a decline in biomass burning (Rice et al. 2016). More recently, since 2006, there have been increases in the atmospheric concentration of methane again (Rigby et al. 2008). It is uncertain whether this is likely to continue or not (Dlugokencky et al. 2009).

Scientists have shown with very high confidence[1] that the increase observed from 1750 to the 1980s was caused by humans (Ciais et al. 2013). Methane emissions

---

[1] This is a term used by the Intergovernmental Panel on Climate Change (IPCC) to denote specific scientific probabilities. The term "very high confidence" conveys a 9/10 chance of being correct. See "Table of Confidence Intervals" in the Glossary of Terms for the full set of IPCC scientific probability terminology.

caused by human activity in 2013 account for between 50% and 65% of total global emissions of methane. Concentrations of methane have been found to be higher downwind of intensive agricultural areas, providing further evidence of the impacts of human methane emissions (Frankenberg et al. 2011).

Like carbon, there are feedback loops associated with the methane cycle. One feedback loop of particular concern pertains to large stores of methane stored in shallow ocean sediments, on the slopes of continental shelves, and in permafrost soils. These are all stable under current temperatures. However, methane emissions contribute to warming which can lead to melting of ice, thus releasing these stores. This would in turn lead to further warming (Ciais et al. 2013).

9.3   Nitrous Oxide

Nitrous oxide (N_2O) is the third most important greenhouse gas. It is a very small component of Earth's atmosphere with a concentration of only 0.33 parts per million parts of atmosphere (ppm). This concentration is 1.2 times higher than the concentration before the Industrial Revolution (Myhre et al. 2013).

Nitrous oxide is released during the natural nitrogen cycle. The largest natural source of nitrous oxide is the soil. Nitrifying and denitrifying bacteria release nitrous oxide as a by-product (see Chap. 14), which accounts for approximately 60% of natural emissions of nitrous oxide (Denman et al. 2007). The remaining emissions come from the oceans (35%) and atmospheric chemical reactions (5%).

Like methane, agriculture is the greatest anthropogenic source of nitrous oxide. Fertilized soil and manure contribute to 42% of methane emissions and runoff and leaching of fertilizers to a further 25%. Biomass burning represents 10%, fossil fuel combustion 10%, biological degradation 9%, and sewage 5% (IPCC 2007).

It can be very difficult to measure nitrous oxide emissions as most agricultural emissions come from bacteria in the soil and because the emissions can vary substantially with weather.

9.3.1   Laughing Gas

Joseph Priestley, an English philosopher, synthesized nitrous oxide in the late 1700s. He called it phlogisticated nitrous air, or inflammable nitrous air. Not long after this, the use of nitrous oxide for medical purposes began. It was used in anaesthetics from the mid-1800s and later as a recreational drug when it was found to put people into a state of euphoric laughter. It is still used in modern medicine[2]

---

[2] Author Kate Meyer was given nitrous oxide to help ease the pain of labour when trying to deliver her first child. To her disappointment, she felt neither euphoric nor any sense of hilarity. She simply felt a bit foggier and slightly ill and is thankful that this was not the only form of pain relief

9.3.2 Impacts

Despite the relatively low concentrations of nitrous oxide in the atmosphere, it is a very dangerous greenhouse gas. One kilogram of N_2O is estimated to cause the same level of warming over 100 years as 298 kg of carbon dioxide or 12 kg of methane.

Moreover, it is now considered to be the single most important ozone-depleting substance. This is not because it has a high impact on ozone depletion per molecule. Nitrous oxide's ozone-depleting potential is relatively low—at only 0.017—whereas CFC-11 (the benchmark for ozone-depleting substances) has an ODP of 1 and some chemicals have ODPs of over 10. However, the low ozone depletion potential is offset by the high quantity of nitrous oxide emissions and its long atmospheric lifetime compared to most ODPs.

N_2O also causes acid rain. Nitrous oxide reacts with water particles to produce nitric acid which has a pH between 4.1 and 5.1. Rain normally has a pH of approximately 5.6. Acid rain can harm plants and other species by dissolving important nutrients and minerals from the soil that are then carried away by the rain. Further, acid rain can lead to the release of toxic substances such as aluminium into soil.

9.4 An Indicator for Me-NO

As mentioned, the warming impacts of GHGs are often equated to an equivalent mass of CO_2 emissions (CO_2e) based on their global warming potential. This indicator corresponds well to the relevant PB indicator—radiative forcing—which is essentially a measure of warming or cooling impacts. Thus, the proposed indicator for methane and nitrous oxide is the *net Me-NO emissions* measured in CO_2e.

9.5 The Limit

There are two Planetary Boundaries which correspond to both methane and nitrous oxide emissions:

- Radiative forcing $\leq \pm 1$ W/m^2
- Extinction rate ≤ 10 extinctions per million species per year

There is a third that relates only to nitrous oxide emissions:

- Concentration of stratospheric ozone ≥ 290 (Dobson units)

As discussed in Chap. 8, the Planetary Boundary for radiative forcing cannot be used to derive a specific limit for any one Planetary Quota. Me-NO emissions

---

available! N_2Ot so funny after all!

contribute to species extinction indirectly through their impacts on climate change. As such, we have assumed that the PB for radiative forcing can be used as a proxy for Me-NO impacts on species extinction rates.

Nitrous oxide is not only a greenhouse gas. It is also an ozone-depleting substance. This means that it reduces the amount of stratospheric ozone—a gas that forms a protective layer that filters ultraviolet radiation from the sun before it reaches Earth's surface. There is no specific limit proposed in the literature for nitrous oxide with respect to stratospheric ozone.

In the earliest publications on Planetary Accounting, the Quota for Ag-GHGs (then referred to as "Me-No" emissions) was ≤5 GtCO$_2$e/year. This was based on the most stringent pathway presented in the fifth Assessment Report by the Intergovernmental Panel on Climate Change (IPCC), *representative concentration pathways* (RCPs) 2.6. RCP2.6 showed end-of-century emissions for methane dropping to ≤143 Mt/year and nitrous oxide to ≤5.3 Mt N/year (Prather et al. 2013). It then assumes constant emissions after 2100 (IPCC 2013a, b). Converting these limits to the control variable unit CO$_2$e (using global warming potential of 198 and 24 for nitrous oxide and methane, respectively) and combing them gave a PQ of *net Ag-GHG emissions* ≤5 GtCO$_2$e/year.

The revised Quota for carbon dioxide (see Chap. 8) is a more stringent final target but allows a higher overall carbon dioxide budget for the remainder of the century. The pathway used as the basis for the Quota for carbon dioxide assumes non-CO$_2$ GHG emissions of 4 GtCO$_2$e in 2100 (Rogelj et al. 2018).

It is not possible to derive limits for specific PQs from the PB for radiative forcing. However, it is possible to meet the Planetary Boundary for radiative forcing using the Ag-GHG limit of 4 GtCO$_2$e. This is demonstrated in Chap. 12.

The layer of ozone in the stratosphere filters harmful ultraviolet rays from the sun and protects life on Earth. In the mid-1970s, scientists realized that some substances could and were depleting the ozone layer (Chesick 1975) (see Chap. 11). There is no specific limit for nitrous oxide or methane within the pathway (Rogelj et al. 2018). However, as the total Ag-GHGs in this pathway are less than the total in RCP2.6, it is assumed that the nitrous oxide share of this budget is no greater than the nitrous oxide share of RCP2.6 which was ≤5.3 Mt N/year. The total ODP of nitrous oxide per year at this level would be 90,100 ODP tonnes (because the ODP of nitrous oxide is 0.017; see Sect. 9.3.2). There has already been a phase out amounting to 2.5 million ODP tonnes of ozone-depleting substances since the Montreal Protocol was ratified (UN 2016). Approximately 32,000 tonnes ODP remain to be phased out. The RCP2.6 target of 5.3 Mt N/year for nitrous oxide (and thus the IPCC1.5 °C pathway discussed above) is considered adequate to respect the PB for ozone on the following basis:

- The authors of the PB framework have indicated that provided the Montreal Protocol commitments are met, the PB for ozone depletion will be respected (Rockström et al. 2009).
- The PB for stratospheric ozone has only been transgressed over Antarctica and only in Australia's spring. Current annual emissions of nitrous oxide are approximately 7.7 Mt N/year. There are a further 32,000 ODPt of Montreal

Table 9.1 Examples of different scales of activity which have or could contribute to achieving the PQ for Me-NO emissions

Achieving the Planetary Quota for Me-NO emissions			
	Community	**Government**	**Business**
Large	Development of an independent global community report on the scientific understanding of Ag-GHGs emissions and corresponding impacts, *e.g. a special report by the Intergovernmental Panel on Climate Change*	Develop a global treaty to limit Ag-GHGs emissions, *e.g. the Paris Agreement*	Develop innovative low/no Ag-GHGs farming practices or solutions
Medium	Campaigns for low Ag-GHG lifestyle choices, *e.g. Veganuary*	Legislate low Ag-GHG farming practices	Develop plant-based meat alternatives, *e.g. the Beyond Meat plant-based burger patties*
Small	Implement low Ag-GHG community projects, *e.g. develop an organic community vegetable garden*	Incentivize low Ag-GHG products, *e.g. organic products*	Go fossil fuel-free or carbon neutral which can include funding agricultural practices
Individual	Eat a low Ag-GHG diet, *e.g. follow the EAT Lancet dietary guidelines which are based on human and planetary health* (Willett et al. 2019)	Drive local awareness around low Ag-GHG behaviours and practices	Educate the public about Ag-GHG emissions, *e.g. Kip Andersen and Keegan Kuhn's documentary Cowspiracy*

gases being emitted. This brings the total emissions of ODPs to approximately 162,900 ODPt. There is evidence that the ozone hole is currently repairing itself despite this residual level of emissions of ozone-depleting substances (Strahan and Douglass 2018). Thus, we are already tracking towards the PB for ozone even in Antarctica with current levels of OPS emissions greater than the proposed PQ levels including nitrous oxide emissions.

On the basis that there is no scientific basis for more stringent limits for nitrous oxide emissions than the RCP2.6 targets, the IPCC1.5 °C pathway limit for Ag-GHGs can be shown to sufficiently respect the corresponding PBs. As such, this limit has been used as the basis for the PQ for Ag-GHGs of *net Ag-GHG emissions* ≤ 4 GtCO$_2$e/year.

9.6   Discussion

The global Ag-GHG footprint is approximately 17.7 GtCO$_2$e/year (derived from (World Bank 2019)), more than four times the annual PQ for Ag-GHGs. Table 9.1 shows examples of how this PQ could be put into practice at different scales of activity and across different sectors. The examples relate to varying timeframes.

9.7 Conclusions

Ag-GHG emissions are considered collectively as they have similar impacts with respect to the Planetary Boundaries. The Planetary Quota for Ag-GHGs is *net Ag-GHG emissions* ≤4 GtCO$_2$e/year. This is based on IPCC1.5 °C P2 pathway targets for 2100.

References

Allan W, Struthers H, Lowe DC (2007) Methane carbon isotope effects caused by atomic chlorine in the marine boundary layer: global model results compared with Southern Hemisphere measurements. J Geophys Res Atmos 112. https://doi.org/10.1029/2006JD007369

Chesick JP (1975) Atmospheric halocarbons and stratospheric ozone. Nature 254:275

Ciais P, Sabine C, Bala G, Bopp L, Brovkin V, Canadell J, Chhabra A, Defries R, Galloway J, Heimann M, Jones C, Le Quéré C, Myneni RB, Piao S, Thornton P (2013) Carbon and other biogeochemical cycles. In: Stocker TF, Qin D, Plattner G-K, Tignor M, Allen SK, Boschung J, Nauels A, Xia Y, Bex V, Midgley PM (eds) Climate change 2013: the physical science basis. Contribution of working group I to the fifth assessment report of the Intergovernmental Panel on Climate Change. Cambridge University Press, Cambridge, UK

Conrad R (1996) Soil microorganisms as controllers of atmospheric trace gases (H$_2$, CO, CH$_4$, OCS, N$_2$O, and NO). Microbiol Rev 60:609–640

Denman KL, Brasseur G, Chidthaisong A, Ciais P, Cox PM, Dickinson RE, Hauglustaine D, Heinze C, Holland E, Jacob D, Lohmann U, Ramachandran S, Da Silva Dias PL, Wofsy SC, Zhang X (2007) Couplings between changes in the climate system and biogeochemistry. In: Climate change 2007: the physical science basis. Contribution of working group I to the fourth assessment report of the Intergovernmental Panel on Climate Change. Cambridge University Press, Cambridge, UK

Dlugokencky EJ, Bruhwiler L, White JWC, Emmons LK, Novelli PC, Montzka SA, Masarie KA, Lang PM, Crotwell AM, Miller JB, Gatti LV (2009) Observational constraints on recent increases in the atmospheric CH$_4$ burden. Geophys Res Lett 36. https://doi.org/10.1029/2009GL039780

Frankenberg C, Aben I, Bergamaschi P, Dlugokencky EJ, van Hees R, Houweling S, van der Meer P, Snel R, Tol P (2011) Global column-averaged methane mixing ratios from 2003 to 2009 as derived from SCIAMACHY: trends and variability. J Geophys Res Atmos 116. https://doi.org/10.1029/2010JD014849

IPCC (2007) Climate change 2007: the physical science basis. Contribution of working group I to the fourth assessment report of the Intergovernmental Panel on Climate Change. Cambridge University Press, Cambridge, UK

IPCC (2013a) Annex II: Climate system scenario tables. In: Stocker TF, Qin D, Plattner G-K, Tignor M, Allen SK, Boschung J, Nauels A, Xia Y, Bex V, Midgley PM (eds) Climate change 2013: the physical science basis. Contribution of working group I to the fifth assessment report of the Intergovernmental Panel on Climate Change. Cambridge University Press, Cambridge, UK

IPCC (2013b) Summary for policymakers. In: Stocker TF, Qin D, Plattner G-K, Tignor M, Allen SK, Boschung J, Nauels A, Xia Y, Bex V, Midgley P (eds) Climate change 2013: the physical science basis. Contribution of working group I to the fifth assessment report of the Intergovernmental Panel on Climate Change. Cambridge University Press, Cambridge, UK

IPCC (2014) Summary for policymakers. In: Field CB, Barros VR, Dokken DJ, Mach KJ, Mastrandrea MD, Bilir TE, Chatterjee M, Ebi KL, Estrada YO, Genova RC, Girma B, Kissel ES, Levy AN, MacCracken S, Mastrandrea PR, White LL (eds) Climate change 2014: impacts, adaptation, and vulnerability. Part A: Global and sectoral aspects. Contribution of working group II to the fifth assessment report of the Intergovernmental Panel on Climate Change. Cambridge University Press, Cambridge, UK

Myhre G, Shindell D, Bréon F-M, Collins W, Fuglestvedt J, Huang J, Koch D, Lamarque J-F, Lee D, Mendoza B, Nakajima T, Robock A, Stephens G, Takemura T, Zhang H (2013) Anthropogenic and natural radiative forcing. In: Stocker TF, Qin D, Plattner G-K, Tignor M, Allen SK, Boschung J, Nauels A, Xia Y, Bex V, Midgley PM (eds) Climate change 2013: the physical science basis. Contribution of working group I to the fifth assessment report of the Intergovernmental Panel on Climate Change. Cambridge University Press, Cambridge, UK

Prather M, Flato G, Friedlingstein P, Jones C, Lamarque JL, Liao H, Rasch P (2013) Annex II: Climate system scenario tables. In: Stocker TF, Qin D, Plattner G-K, Tignor M, Allen SK, Boschung J, Nauels A, Xia Y, Bex V, Midgley PM (eds) Climate change 2013: the physical science basis. Contribution of working group I to the fifth assessment report of the Intergovernmental Panel on Climate Change. Cambridge University Press, Cambridge, UK

Rice AL, Butenhoff CL, Teama DG, Röger FH, Khalil MAK, Rasmussen RA (2016) Atmospheric methane isotopic record favors fossil sources flat in 1980s and 1990s with recent increase. Proc Natl Acad Sci U S A 113:10791–10796

Rigby M, Prinn RG, Fraser PJ, Simmonds PG, Langenfelds RL, Huang J, Cunnold DM, Steele LP, Krummel PB, Weiss RF, Doherty S, Salameh PK, Wang HJ, Harth CM, Mühle J, Porter LW (2008) Renewed growth of atmospheric methane. Geophys Res Lett 35. https://doi.org/10.1029/2008GL036037

Rockström J, Steffen W, Noone K, Persson A, Chapin FS, Lambin E, Lenton TM, Scheffer M, Folke C, Schellnhuber HJ, Nykvist B, De Wit CA, Hughes T, van der Leeuw S, Rodhe H, Sorlin S, Snyder PK, Costanza R, Svedin U, Falkenmark M, Karlberg L, Corell RW, Fabry VJ, Hansen J, Walker B, Liverman D, Richardson K, Crutzen P, Foley J (2009) Planetary Boundaries: exploring the safe operating space for humanity. Ecol Soc 14:32

Rogelj J, Shindell D, Jiang K, Fifita S, Forster P, Ginzburg V, Handa C, Kheshgi H, Kobayashi S, Kriegler E, Mundaca L, Séférian R, Vilariño MV (2018) Mitigation pathways compatible with 1.5 °C in the context of sustainable development. In: Masson-Delmotte V, Zhai P, Pörtner HO, Roberts D, Skea J, Shukla PR, Pirani A, Moufouma-Okia W, Péan C, Pidcock R, Connors S, Matthews JBR, Chen Y, Zhou X, Gomis MI, Lonnoy E, Maycock T, Tignor M, Waterfield T (eds) Global warming of 1.5 °C. An IPCC special report on the impacts of global warming of 1.5 °C above pre-industrial levels and related global greenhouse gas emission pathways, in the context of strengthening the global response to the threat of climate change, sustainable development, and efforts to eradicate poverty. IPCC, Cambridge, UK

Strahan SE, Douglass AR (2018) Decline in Antarctic ozone depletion and lower stratospheric chlorine determined from aura microwave limb sounder observations. Geophys Res Lett 45:382–390

UN (2016) Montreal protocol - submission to the high-level political forum on sustainable development (HLPF) 2016. Geneva, United Nations Sustainable Development Knowledge Platform: UN

Willett W, Rockstrom J, Loken B, Springmann M, Lang T, Vermeulen S, Garnett T, Tilman D, Declerck F, Wood A, Jonell M, Clark M, Gordon LJ, Fanzo J, Hawkes C, Zurayk R, Rivera JA, De Vries W, Majele Sibanda L, Afshin A, Chaudhary A, Herrero M, Agustina R, Branca F, Lartey A, Fan S, Crona B, Fox E, Bignet V, Troell M, Lindahl T, Singh S, Cornell SE, Srinath Reddy K, Narain S, Nishtar S, Murray CJL (2019) Food in the Anthropocene: the EAT-Lancet Commission on healthy diets from sustainable food systems. Lancet 393:447–492

World Bank (2019) CO_2 emissions (metric tons per capita). http://data.worldbank.org/indicator/EN.ATM.CO2E.PC?order=wbapi_data_value_2009%20wbapi_data_value%20wbapi_data_value-last&sort=asc. Accessed 18 May 2019

Chapter 10
A Quota for Forestland

The one who plants trees, knowing that he will never sit in their
shade, has at least started to understand the meaning of life.
Rabindranath Tagore

Abstract Humans have been altering Earth's surface for more than 40,000 years. Deforestation and land-use change have occurred for much longer than many of the more "modern" impacts that have been occurring predominantly since the Industrial Revolution. All the same, land-use change has accelerated since the Industrial Revolution which has local and global impacts.

Forests play critical roles in the maintenance of the state of the Earth system. They are an integral part of the carbon, water, and nitrogen cycles. They provide important habitats. Moreover, they provide important resources for humans such as timber and food. The Planetary Boundary for land-use change is for forest area ≥75% of original forest area. Only 62% of original forest area is still forest now.

Forest area is also critical to several other Planetary Boundaries. The total area of global forest effects the Planetary Boundaries for land-use change, climate change, and biosphere integrity.

The Planetary Quota for forested land is *net deforestation* ≤−11 Mha/year. This can be compared with the net reforestation or deforestation associated with any scale of human activity. The limit is set on the basis of meeting the Planetary Boundaries for land-use change, climate change, and biosphere integrity by the end of this century.

10.1 Introduction

Earth's surface is 510 million km². Seventy-one percent of the surface is covered by oceans, leaving only 29%, or approximately 150 million km² of land. Of this land area, almost one third is classified as desert—defined as areas that have less total rainfall than evaporation over a year. Deserts have harsh conditions and are typically only very scarcely populated by living creatures or plants. This means that there are approximately 100 million km^2 available to support most terrestrial species, including humans.

Four land-use-based pressures were found to be critical pressures with respect to the Planetary Boundaries:

- Deforestation
- Land conversion
- Land segregation
- Land degradation

All four pressures affect the Planetary Boundaries for radiative forcing, aerosol optical depth, extinction rate, and remaining forest area. Deforestation also effects the Planetary Boundary for the concentration of CO_2 in the atmosphere. It is one of the key mechanisms that will enable us to return to and operate within this limit. As such, there are two land-use-related Planetary Quotas, a Planetary Quota for forestland (discussed in this chapter) and a Planetary Quota for biodiversity loss—which uses a land-use proxy indicator (discussed in Chap. 16).

This chapter begins with a background of human manipulation and management of land and an overview of why forestland is of particular importance. This is followed by the case for the indicator selected and the scientific basis for the proposed limit. The chapter concludes with a discussion about the PQ for reforestation in the context of today's deforestation practices.

10.2   Background

Intentional changes to natural landscape by humans can be traced back as far as 40,000 years to Australia with some evidence that it could be 60,000 years. Fire is a natural part of the Australian landscape. Aboriginal people learnt to use it to their advantage. Nyungar people—Aboriginals from the south-west of Western Australia—used "cool" and "hot" fires, fires of low and high intensity, respectively—for different purposes. Cool fires were used to clear undergrowth for better access through dense bush and to promote new growth that would attract animals that could then be farmed; plant species with high nutritional value were the first to re-establish themselves after a cool burn. Cool fires were also used to promote the growth of grass. Fires were (and still are) used to maintain grazing habitats. "Hot" fires, which burn not only the undergrowth but also the middle and upper layers of forest or bush, were used to promote new grown wattan or spearwood thickets (Kelly 1999) which had other uses in terms of spears and other implements.

Not all other cultures managed land-use changes to their advantage. Deforestation has been a primary cause of several societal collapses.

When the first Polynesians arrived at Easter Island in approximately 800 AD, it was covered in tropical forest, with huge palm trees, and dandelions as tall as trees. The island was home to the largest collection of breeding sea-birds in the Pacific. Over the years they cleared

Fig. 10.1 Easter Island statues against a desolate backdrop void of trees. (Massardier 1998, CC BY-SA 3.0 (CC BY-SA 3.0: Creative Commons License allows reuse with appropriate credit))

forestland for houses and gardens, and used timber for canoes, firewood, and to transport and lever into place their giant statues. By 1600 there were no trees left (see Fig. 10.1). They had also hunted all but one of the sea-birds to extinction. Without forest, they had no fruit, no timber for canoes to go fishing, no fuel for fires. Without the tree roots, the land eroded quickly, and agricultural yields dropped. The most widely available food remaining was themselves, and so they turned to cannibalism. The society collapsed due to this sequence of land use mismanagement (Diamond 2005).

The Anasazi collapse can also be attributed to deforestation. The Anasazi were Native-American people who were hugely advanced in many ways for their time. For example, they constructed buildings as high as 6 storeys with as many as 600 rooms beginning around 600 AD. However, they did not have good forest-management practices. They cleared the forests close to their settlements, and then in a slightly broader radius, until the point that they were travelling 75 miles, to mountains 4000 feet above their settlement for timber to use as fuel and construction materials. This timber all needed to be dragged back to the settlements by hand. The Anasazi survived several droughts by relocating their settlements, but in 1117, another drought occurred and there was no unexploited landscape left. This society collapsed two decades later. The environment remains void of trees (Diamond 2005).

From 27 BC the Roman Empire prospered for almost 500 years. Yet in 476, the last emperor was removed from power, and the Roman Empire collapsed. Deforestation is now believed to be one of the primary causes of this sudden collapse. Wood was an important resource used for building, heating, for fire in industry and for their war machines such as boats and chariots. Wood was overharvested from some forest areas, while others were cleared entirely to make room for farmland to feed the rapidly growing population. Forests were also burnt down in response to native tribes who would escape into forests to launch surprise attacks. One of the greatest impacts of the deforestation was the loss of topsoil from hillsides to lowland areas. The hillsides were no longer productive and crop yields dropped. The erosion led to silting of rivers which interfered with cities and their economies, eg Ephesus, the second biggest city in the Roman empire was abandoned after the river it was on silted up and its function as a port became impossible. Increased flooding from siltation led to the lowlands forming marshes that were breeding areas for disease. In light of the lack of fuel, glass and brick industries closed down or relocated. The Roman Empire gradually weakened and was eventually overrun with Barbarians (Sing 2001).

The active management of forests, not including the controlled burning by Aboriginals, can be traced back as early as the 1400s to Venice. Wood was the essential foundation of all wealth at this time as without wood there were no ships, which meant no trade, defence, or power. The Venetian Great Council wrote laws to attempt to ensure adequate supplies of wood would be available for ship building (Mauch 2013). However, these laws were targeted at reducing demand and not managing supply. For example, boat captains were charged fines for damaging oars. The laws were unsuccessful but give an interesting insight into the development of our understanding of forest systems. In the 1400s in Venice, there were already elaborate mapping and measuring methods in place. In the sixteenth century, time and not just space began to be incorporated into forest planning (Mauch 2013). Silviculture—the practice of managing and maintaining forest systems—was first introduced in the 1700s by von Carlowitz (1713).

Despite these examples of our very early understanding of forest management and the importance of this, we are continuing to use forest resources faster than they can regenerate across the globe. Before the industrial era, there were approximately 59 million km^2 of forest area. Today, there are less than 40 million km^2 remaining.

The Amazon Rainforest is the world' largest tropical rainforest. Since the 1970s, nearly 20% of the 4000,000 km^2 rainforest has now been deforested. Despite the critical roles this forest plays in Earth's carbon and hydrological cycles, deforestation has hit record highs in 2019 with monthly deforestation up 88.4% in June 2019 compared with June 2018. It has been estimated that the Amazon generates approximately half of its own rainfall. At a certain level of deforestation, the forest will thus no longer be able to sustain itself, at which point the rainforest would degrade to a savannah. It has been estimated that the tipping point for the Amazon is between 20% and 25% deforestation (Lovejoy and Nobre 2018).

Human impacts on forestland are not limited to chopping them down. Overhunting can lead to empty forest syndrome where ecosystems can collapse because of the lack of a key species. The demolition of relatively small areas of large forests, for example, to run roads through them, can cause fragmentation—harming the forest function. Climate change has altered natural fire regimes that are fundamental to forest health. The introduction of alien species can damage forests. Air pollution can lead to acid rain which damages forests.

10.3   The Importance of Forests

More than half of the land surface area on Earth is classified as "arid". This includes cold regions (polar and tundra areas as well as high mountains and plateaus) which comprise 14% of the global land area. Drylands—including hyper-arid zones (true deserts), arid zones (less than 200 mm annual rainfall), semiarid zones (seasonal rainfall regimes with max rainfall of 800 mm), and dry subhumid zones (highly seasonal rainfall)—comprise a further 47% (Secretariat of the CBD 2001).

Drylands are used for farming both crops and livestock. Grazing is a major use of global drylands. Drylands are at higher than normal risk of erosion. Poor management of nondesert drylands can and has led to desertification—almost 70% of global dryland area, 35 million km², is already affected by desertification (Secretariat of the CBD 2001). Many dryland ecosystems sustain a degree of natural fire. However, human-caused fires can still have impacts on the biodiversity of these ecosystems.

The nondesert surface area of Earth can be roughly categorized into forest, cropland, urban areas, and drylands. One of the key human pressures on biodiversity is habitat loss due to human activity, i.e., land-use change. Forests are the most important habitat as approximately 80% of the world's terrestrial species are found in tropical rainforests and even localized deforestation can lead to the extinction of important species (WWF 2014). Forests also provide a lot of critical natural capital to humans, they are an important part of the carbon cycle, and they are our supply for timber, paper, fuel, and other wood-based products.

Not all forests are equal. Tropical forests are the most diverse ecosystems on Earth. Forest plantations on the other hand, which cover more than 1 million km² of Earth's surface, are usually made up of a single tree species—frequently an introduced species. These forests are not a popular habitat and usually have low levels of biodiversity. There are management practices that can encourage species diversity. For the purpose of this thesis, forest area is defined as per the Kyoto definition (see Box 10.1).

Deforestation affects climate change in two key ways. It is responsible for as much as 17% of global emissions of carbon dioxide (IPCC 2007). Deforestation also affects the reflectivity of Earth's surface. As much as -0.15 W/m² $\pm$ 1 of radiative forcing (change in energy balance at Earth's surface) is attributed to changes in surface albedo.

Deforestation can also lead to aerosol pollution. Aerosols are small particles suspended in the atmosphere. They come from both natural and anthropogenic sources. One of the most common naturally occurring aerosols is atmospheric dust particles. Deforestation can lead to erosion and eventually desertification. As more areas become desert, there is more loose sand and dust that can be carried into the atmosphere.

Box 10.1 Defining Forest

There is no singular definition for "forest area" that is widely accepted. The Convention on Biological Diversity (CBD) defines forests loosely as "ecosystems in which trees are the predominant life forms" (Secretariat of the CBD 2001). The Food and Agriculture Organization (FAO) of the United Nations defines a forest as "spanning more than 0.5 hectares with trees higher than 5 metres and a canopy cover of more than 10 percent, or trees being able to reach these thresholds in situ" (FAO 2012). FAO do not include land that is predominantly used for agriculture or urban land as forest area. However, the CBD find the FAO definition very broad and suggest that a more rigorous definition for forest would be that of closed canopy forest. Yet even the definition for closed canopy forest ranges from thresholds of 30–70% canopy cover. The Kyoto definition is similar to the FAO definition but only requires 10–30% of crown cover with the potential to meet 2–5 m in height at maturity (UU 1998). This definition continues, to clarify that forests can consist of closed formations or open forests. They include young forests which have not yet reached the required crown density and temporarily unstocked forests (whether this is from human or natural causes) provided these are expected to revert to forests.

Defining forest is important. Some tree species such as the Australian native, Mallee, do not fit traditional European-based definitions of forest as they are multi-stemmed, yet they withdraw carbon, prevent erosion, and provide natural habitat for local species. They have deep roots that can store carbon for several hundreds of years and provide effective agro-forestry opportunities with shelter and shade for sheep producing more productive outcomes than just having grass. The Kyoto definition is the broadest of the formal definitions and as such is the definition used in this thesis though this is likely to be changed by IPCC processes.

10.4   An Indicator for Forestland

The original PB indicator for land-use change was the percentage of global ice-free land surface converted to cropland (Rockström et al. 2009). This was updated in 2015 to be forest area ≥75% of original forest area (Steffen et al. 2015). The updated limit was based on the regulation of the climate system and the hydrological cycle. The current status of global forested land is 62% of original—i.e., we have exceeded this boundary (Steffen et al. 2015).

Deforestation affects up to four of the PB limits:

- Remaining forest ≥75% original forest area
- Extinction rate ≤10 extinctions per million species per year
- Radiative forcing ≤1 W/m^2
- Aerosol optical depth ≤0.1

Given that the area of forest needed (i.e., 75%) is more than the current area of forest (i.e., 62%), the limit for deforestation will be negative, i.e., a minimum rate of reforestation in units of hectares will be needed.

10.5 The Limit

It is difficult to determine baseline levels for forest area as human destruction of forest started approximately 10,000 years ago at the beginning of the Holocene over most of the world, and even earlier in Australia. It is thought that prior to the industrial era there was approximately 59 million km^2 of forest area. Before human influence it is estimated that about half of Earth's surface was forest or woodland—i.e., approximately 75 million km^2 (Secretariat of the CBD 2006). Of the world's 15 billion hectare of surface area, only 6.5 billion hectare of this is suitable for forestry. This figure is taken to be the baseline figure for "original forest". Table 10.1 shows estimates for minimum forest areas based on each of the PBs which corresponds with the pressure deforestation.

The minimum forest area needed to meet the PB for land-use change is roughly equal to the upper estimate for the forest area needed to meet the PB for the concentration of CO_2 in the atmosphere. As such, the PQ for reforestation is deforestation ≤ -11 Mha/year. This can be compared to net deforestation of any scale of human activity.

Table 10.1 Summary of global deforestation limits based on different upstream PBs

Planetary Boundary	Minimum area to be reforested	Basis
Land-use change >75% original forest area restored	≤ -0.9 Gha or ≤ -11 Mha/year	75% of original forest area equates to total forest area of 4.9 billion hectares. Current forest area is approximately 4 billion hectares
Climate change: CO_2 concentration ≤ 350 ppm	≤ -0.9 Gha or ≤ -11 Mha/year	This would require reforestation between 0.6 (Watson et al. 2000) and 0.9 (Brown et al. 1996) billion hectare by 2100. At the high end of this range, this equates to total forest area of 4.9 billion hectare
Climate change: radiative forcing ≤ 1 W/m^2	NA	The PB for radiative forcing cannot be used to derive a minimum forest area. However radiative forcing impacts of deforestation and land-use change are discussed in Chap. 12
Biosphere integrity: Extinction rate ≤ 10 E/MSY	NA	There is no specific global forest area that relates to extinction rate. There are estimations of how much land should be retained for biosphere integrity—the "biodiversity buffer" which ranges from 1% to 99% (Fahrig 2001; Wackernagel et al. 2002; The Brundtland Commission 1987; Margules et al. 1988). There is a land-based PQ for biodiversity (see Chap. 16). As such, the limits pertaining to the PBs for land-use and climate change are considered adequate

10.6   Discussion

From 2010 to 2015, the average rate of deforestation was 6.5 Mha/year (FAO 2016). To reverse this and achieve a reforestation rate of 11 Mha/year will be challenging.

Table 10.2 lists examples of some of the different activities that could occur at different scales to help to make this transition.

10.7   Conclusions

Forest is the most critical land type for healthy Earth-system functioning. This is because of its role in the climate system, and on the global water cycle, and because it provides important habitats to so many species as well as providing critical resources for human habitation.

Currently we are destroying forests at a rate of approximately 6.5 Mha/year. Only 62% of original forest remains. The Planetary Quota for forestland is *net deforestation* ≤ -11 Mha/year. This can be compared to the net deforestation associated with any scale of human activity.

Table 10.2 Examples of different scales of activity which have or could contribute to achieving the PQ for forestland

Achieving the Planetary Quota for forestland			
	Community	**Government**	**Business**
Large	Develop global sustainable forestry standards, e.g., *the Forest Stewardship Council*	Develop a global treaty on forestry	Reduce the footprint of business operations and reforest previously occupied land
Medium	Run tree planting community events	Legislate minimum national forest cover, e.g., *Bhutan has a law that at least 60% of the country must be under forest cover*	Innovate to incorporate increasing forest area in business practices,e.g., *CapitaLand have built an office tower "Capita Green" in Singapore with 50 times greater volume of forest than the building footprint (i.e., the land area within the building perimeter) would have had as original forest area through innovative biophilic design*
Small	Community tree planting	Run tree planting community events	Only purchase sustainably sourced forestry products
Individual	Plant trees		Go carbon neutral and use the carbon deficit to fund the planting of trees

References

Brown S, Sathaye J, Cannel M, Kauppi P (1996) Management of forests for mitigation of greenhouse gas emissions. In: Watson RT, Zinyowera MC, Moss RH (eds) Climate change 1995: impacts, adaptations, and mitigation of climate change: Scientific-technical analyses, Contribution of working group II to the second assessment report of the Intergovernmental Panel on Climate Change. Cambridge University Press, Cambridge, UK

Diamond JM (2005) Collapse: how societies choose to fail or succeed. Viking, New York

Fahrig L (2001) How much habitat is enough? Biol Conserv 100:65–74

FAO (2012) FRA 2015 terms and definitions. FAO, Rome

FAO (2016) Global Forest Resources Assessment 2015. FAO, Rome

IPCC (2007) Climate change 2007: the physical science basis. Contribution of working group I to the fourth assessment report of the Intergovernmental Panel on Climate Change. Cambridge University Press, Cambridge, UK

Kelly G (1999) Karla Wongi: fire talk. Landscope 14(2):48–53

Lovejoy TE, Nobre C (2018) Amazon tipping point. Sci Adv 4:eaat2340

Margules CR, Nicholls AO, Pressey RL (1988) Selecting networks of reserves to maximise biological diversity. Biol Conserv 43:63–76

Massardier G (1998) Moais_Anakena.job. Wikimedia

Mauch C (2013) The growth of trees—a historical perspective on sustainability, Carl-von-Carlowitz Series. Oekom, Munich

Rockström J, Steffen W, Noone K, Persson A, Chapin FS, Lambin E, Lenton TM, Scheffer M, Folke C, Schellnhuber HJ, Nykvist B, De Wit CA, Hughes T, van der Leeuw S, Rodhe H, Sorlin S, Snyder PK, Costanza R, Svedin U, Falkenmark M, Karlberg L, Corell RW, Fabry VJ, Hansen J, Walker B, Liverman D, Richardson K, Crutzen P, Foley J (2009) Planetary boundaries: exploring the safe operating space for humanity. Ecol Soc 14:32

Secretariat of the CBD (2001) Global biodiversity outlook 1. UNEP, CBD, Montreal

Secretariat of the CBD (2006) Global biodiversity outlook 2. UNEP, CBD, Montreal

Sing C (2001) World ecological degradation: accumulation, urbanization, and deforestation, 3000 B.C.–A.D. 2000. Rowman Altamira, Lanham

Steffen W, Richardson K, Rockström J, Cornell SE, Fetzer I, Bennett EM, Biggs R, Carpenter SR, De Vries W, De Wit CA, Folke C, Gerten D, Heinke J, Mace GM, Persson LM, Ramanathan V, Reyers B, Sörlin S (2015) Planetary boundaries: guiding human development on a changing planet. Science 347:1259855

The Brundtland Commission (1987) Our common future. World Commission on Environment and Development, Geneva

UN (1998) Kyoto protocol to the United Nations framework convention on climate change. UN Doc FCCC/CP/1997/7/Add.1, Dec. 10, 1997; 37 ILM 22

von Carlowitz HC (1713) Sylvicultura oeconomica, oder haußwirthliche Nachricht und Naturgemäße Anweisung zur Wilden aum-Zucht. Verlag Kessel, RemagenOberwinter, Germany

Wackernagel M, Schulz NB, Deumling D, Linares AC, Jenkins M, Kapos V, Monfreda C, Loh J, Myers N, Norgaard R, Randers J (2002) Tracking the ecological overshoot of the human economy. Proc Natl Acad Sci U S A 99:9266–9271

Watson RT, Noble IR, Bolin B, Ravindranath NH, Verardo DJ, Dokken DJ (2000) Land use, land-use change, and forestry. IPCC, Cambridge, UK

WWF, WWL (2014) Deforestation. http://worldwildlife.org/threats/deforestation. Accessed 27 Mar 2014

Chapter 11
A Quota for Ozone-Depleting Substances

Perhaps the single most successful international agreement to date has been the Montreal Protocol
Kofi Annan

Abstract A thin layer of ozone in the atmosphere protects humans and terrestrial life from harmful ultraviolet radiation. The human creation and emission of ozone-depleting substances has thinned this layer so much that once a year a large localized region with almost no ozone appears over Antarctica. This is known as the hole in the ozone layer.

In 1989, a global treaty was put into place to ban the manufacture and use of substances which deplete the ozone layer, the Montreal Protocol. By 2009 the protocol had been ratified by every country. There has been a reduction of almost 98% in the use of ozone-depleting substances. The ozone hole is starting to get smaller.

The hole in the ozone layer is an example of how human activity can alter global Earth-system processes. It is also an example of how the global population can work together to begin to repair past environmental damage. It is thought that provided we respect the terms of the Montreal Protocol, the hole will repair itself before the end of this century. Thus, the Planetary Quota for ozone is zero emission of ozone-depleting substances listed under the Montreal Protocol. Ozone-depleting substances can be converted to a unit of "ozone-depleting potential" tonnes. This unit can be used to compare emissions of Montreal gases of any scale of human activity to the Planetary Quota limit.

11.1 Introduction

The atmosphere is divided into five primary layers (Fig. 11.1). Some of these layers have secondary layers within them. The troposphere is the layer closest to Earth's surface, the layer in which we live. The second lowest layer in the atmosphere is called the stratosphere. The stratosphere is more than 10 km above Earth's surface (Fahey 2003).

© Springer Nature Singapore Pte Ltd. 2020
K. Meyer, P. Newman, *Planetary Accounting*,
https://doi.org/10.1007/978-981-15-1443-2_11

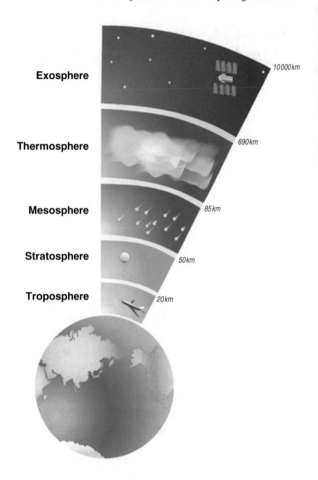

Fig. 11.1 The five layers of the atmosphere. (Unknown 2015, CC BY-SA 4.0 (CC BY-SA 4.0: Creative Commons License allows reuse with appropriate credit))

At the bottom of the stratosphere is a very thin layer which contains relatively high concentrations of ozone (O_3), of up to 12 parts per million (ppm) (Fahey 2003). This layer is called the ozone layer. The ozone layer is important as it protects life on Earth by filtering most of the ultraviolet radiation from the sun before it reaches Earth's surface.

Human emissions of ozone-depleting substances (ODSs), for example, refrigerants such as CFCs, have reduced the concentration of ozone in the ozone layer. The thinning of ozone is large and localized over Antarctica. Each year, there is a period when there is almost no ozone in this area. This has been labelled the "hole" in the ozone layer.

The story of the hole in the ozone layer is one that conveys how large the consequences of human activity can be on the functioning of the Earth system. However, it can also be viewed as a success story, a way of showing how the global community can work together to successfully manage the Earth system and move away from dangerous environmental tipping points. In 1987, in response to the hole in the ozone layer, world leaders agreed on the Montreal Protocol, a global accord to phase out key ODSs (UNEP 2017a). The ODSs included under the protocol are now known as "Montreal gases". This phase out, which will not be complete until 2030, has already led to a decrease in the rate of ozone depletion over Antarctica (Strahan and Douglass 2018). Some scientists estimate that the hole could be closed by 2050 (e.g., Solomon et al. 2016). Others predict that even by 2080, there may still be a small hole, but that we are making progress in the right direction (Strahan and Douglass 2018).

The critical pressures that influence the Planetary Boundary for ozone depletion are emissions of:

- Halocarbons
- Hydrochlorofluorocarbons
- Halons
- Chlorofluorocarbons
- Carbon tetrachloride
- Methyl chloroform
- Hydrochlorofluorocarbons
- Methyl bromide
- Bromochloromethane
- Hydrobromofluorocarbons
- Hydrofluorocarbons
- Nitrous oxide

All but nitrous oxide are covered under the Montreal Protocol and addressed through the Planetary Quota for ozone-depleting substances. We refer to these gases as the "Montreal gases" herein.

The chapter begins with an overview of human influence on the ozone layer. This is followed by a discussion about the Montreal Protocol and an overview of how ODSs are currently being managed. The case for the PQ indicator and limit is then presented and the exclusion of nitrous oxide from this Quota discussed. The chapter concludes with a discussion about current Montreal gas emissions and potential actions that will still be needed to phase these out.

11.2 Background

Ozone (O_3) is a gas that occurs naturally in the atmosphere through ultraviolet sunlight reactions with oxygen molecules. The more sunlight, the more reactions, and so the ozone layer is thickest in the tropics (where it is most needed). When people claim that they can smell oncoming rain, this may well be the scent of ozone. Ozone is so smelly that it can be detected even in very low concentrations. Indeed, the name ozone comes from the Greek word ozein which means "to smell" (Fahey 2003). Lightening can split nitrogen and oxygen which can lead to the creation of ozone, so those down-wind of the lightning may well be able to smell that the storm is approaching.

There is very little ozone in the atmosphere. If all the ozone molecules were collected and distributed across Earth's surface, the layer of pure O_3 gas would be less than half a centimetre (Fahey 2003). Ozone is measured in Dobson units (DU), a measure of the amount of a gas in a vertical column of atmosphere, with total ozone values varying between 200 and 500 (DU) across the globe. Ozone can be depleted naturally through reactions with naturally occurring chemicals. The total abundance of ozone at a given time is determined by the rate of production and the rate of depletion (Fahey 2003).

Approximately 90% of all atmospheric ozone is in the stratosphere (Fahey 2003). Ozone protects life on Earth by filtering harmful wavelengths of ultraviolet (UV) rays from the sun. There are three categories of UV radiation, UV-A, UV-B, and UV-C. UV-C is the most harmful. This is entirely screened out by dioxygen (for wavelengths less than 200 nm) and ozone (for wavelengths above 200 nm). UV-B radiation is less harmful than UV-A but sill increases the risk of skin cancer, cataracts, and the suppression of the immune system (Fahey 2003). It can also harm plant life, single-cell organisms, and marine organisms (Fahey 2003). Most UV-B is filtered out by the ozone layer. UV-A radiation still reaches Earth's surface. It can still damage the skin, but it is far less harmful than shorter wavelength radiation (Fahey 2003).

The remaining 10% of ozone in the atmosphere is located in the troposphere—the closest layer of the atmosphere to Earth's surface. Tropospheric ozone is a greenhouse gas. Ozone in the troposphere can reduce crop yields and forest growth. Humans exposed to ozone can have reduced lung capacity, chest pains, throat irritations, and coughing; it can also worsen pre-existing heart and lung conditions (Fahey 2003). Tropospheric ozone is naturally occurring and performs important functions such as the removal of methane, carbon monoxide, and nitrogen oxides. However, the increased levels from human activity have negative consequences. This is particularly true when increased concentrations of ozone are near humans, plants, and animals.

Ozone can be depleted by free radicals including nitric oxide, nitrous oxide, hydroxyl, chlorine, and bromine. In the 1980s man-made compounds such as chlorofluorocarbons and bromofluorocarbons in common household and industrial products led to a substantial increase of chlorine and bromine radicals in the atmosphere—the foundations of most ozone-depleting substances. These increases in chlorine and bromine led to rapid depletion of stratospheric ozone. The largest hole in the ozone, where depletion has led to almost no ozone in a large localized area, is over Antarctica. This is in part because the ozone layer is thinner at the poles and because the ozone-depleting substances are predominantly released upwind of the Antarctic.

It was first recognized that some substances could deplete and were depleting the ozone layer in the mid-1970s (Chesick 1975). In 1985, an article was published in *Nature* confirming that there was a repeating springtime hole in the ozone layer (Farman et al. 1985). In response, the Montreal Protocol was developed.

11.2.1 *Montreal Protocol*

The Montreal Protocol is a global treaty put into place to manage human impacts on the ozone layer. It was first ratified in 1989 (UNEP 2017a). By 2009 it became the first treaty to have universal ratification (UNEP 2017b).

The Montreal protocol comprises a staged phaseout of "Montreal gases". There are different phaseout dates for different gases and for countries of varying wealth. The gases included in the Montreal Protocol and their phaseout dates are detailed in Table 11.1.

11.2.1.1 The Kigali Amendment

Hydrofluorocarbons (HFCs) were developed in the 1980s as a replacement to ozone-depleting substances. It was later discovered that this substance is a very dangerous greenhouse gas with warming impacts thousands of times higher than CO_2 per unit of emission. Although the concern relating to this substance is for climate change rather than ozone depletion, the management of HFCs has fallen under the Montreal Protocol.

11.2.1.2 Nitrous Oxide

In the past, not much consideration was given to nitrous oxide's impacts on the ozone layer despite the knowledge that it is an ozone-depleting substance. The reason for this is the gas's relatively low ozone depletion potential (ODP). The ODP of a substance is a measure of the impacts of a kilogram of emissions with respect to ozone, compared to the impacts of CFC-11. CFC-11 is considered the benchmark gas with an ODP of 1. Some substances have much higher ODPs than CFC-11, for example, bromochlorodifluoromethane has an ODP of 7.9. Nitrous oxide has an ODP of 0.017.

Despite the low relative impacts of nitrous oxide, it has recently been brought under the spotlight as one of the most important ozone-depleting substances. This is because of its long atmospheric lifetime and the high and growing level of anthropogenic nitrous oxide emissions. Nitrous oxide is not currently included under the Montreal Protocol.

The basis for the exclusion of nitrous oxide from the PQ for ozone is not its omission from the Montreal Protocol. The reason is that the limit for Montreal gases—both for the Montreal Protocol and for the Planetary Quota—is zero.

Nitrous oxide is a naturally occurring by-product of agriculture. A limit of zero is not currently conceivable without putting an end to agriculture. To include it here would mean the limit for this PQ would have to be higher than zero. As such, nitrous oxide emissions are dealt with separately under the PQ for Me-NO. Chapter 9 shows

Table 11.1 Montreal gases: uses and phaseout dates

Montreal gas	Description	Phaseout
Halons	Any group of organohalogen compound containing bromine or fluorine and one or two carbons. They are predominantly used to extinguish fires—they were very useful because they do not conduct electricity and could thus be used to put out electrical fires	Developed Countries 1993 Developing Countries 2010
Chlorofluorocarbons	These are most notable for their use to replace toxic refrigerants ammonia, methyl chloride, and sulphur dioxide which led to fatal incidents in the 1920s due to refrigerant leaks. CFCs were invented in 1928 by Thomas Midgley of General Motors, and in 1930 General Motors and DuPont formed a company to produce CFCs (called Freon) in large quantities. CFCs were also used in aerosol sprays, blowing agents for foams and packing materials, and as solvents	Developed Countries 1995 Developing Countries 2010
Carbon tetrachloride	This was used to produce chlorofluorocarbon refrigerants. It was also used in lava lamps, by stamp collectors to reveal watermarks, as a solvent, a cleaning agent, and in fire extinguishers. It was first made by Henri Victor Regnault in 1839	Developed Countries 1995 Developing Countries 2010
Methyl chloroform	This was also developed by Henri Victor Regnault in 1840 and was used as a solvent	Developed Countries 1995 Developing Countries 2015
Hydrobromofluorocarbons	These were used in Canada for experimental purposes but were identified as ozone-depleting substances and phased out before they became produced or used commercially	Developed Countries 1995 Developing Countries 1995
Methyl bromide	In 1999, an estimated 71,500 tonnes of synthetic methyl bromide were used annually worldwide (UNEP 1999). Almost all of this was for fumigation	Developed Countries 2005 Developing Countries 2015
Bromochloromethane	This was invented in Germany in the mid-1940s as a less toxic fire extinguisher to carbon tetrachloride	Developed Countries 2002 Developing Countries 2002

(continued)

Table 11.1 (continued)

Montreal gas	Description	Phaseout
Hydrochlorofluorocarbons	These are similar to CFCs but with less impact on the ozone layer—they have been used to replace CFCs as refrigerants. They are also used in insulative foams	Developed Countries 2020 Developing Countries 2030
Hydrofluorocarbons	These are often used as refrigerants. They do not harm the ozone layer as much as the refrigerants they are replacing, but they are a dangerous greenhouse gas. They are also used as blowing agents, extinguishers, cleaning products, and propellants. These substances were not originally included in the Montreal Protocol; however on October 15, 2016, an amendment was adopted to phase these down by more than 80%	Developed Countries (85% Reduction) 2035 Developing Countries (80% reduction) 2045

that the impacts to the ozone layer from the proposed limit for nitrous oxide are unlikely to prevent the recovery of the ozone layer.

11.3   The Limit

Ozone-depleting potential of substances can be used to compare different quantities of emissions of each using the unit ODPkg. The PQ indicator for ozone is thus *emissions of Montreal gases* in ODPkg.

The Planetary Boundary for ozone depletion is ≤5% decrease in column ozone levels for any latitude with respect to 1964–1980 values, a concentration of stratospheric ozone ≥290 DU (Chipperfield et al. 2006). It is difficult to equate this limit to an emissions budget. However, in the Planetary Boundaries, publication Rockström et al. (2009) state that the Montreal Protocol has put humanity on a path that will avoid the transgression of the PB for ozone, citing evidence of a decrease of 8–9% by 2005 of tropospheric concentrations of ozone-depleting gases from their peak values in 1992–1994 (Clerbaux et al. 2006).

The Montreal Protocol comprises a complete phaseout of Montreal gases other than HFCs by 2030. Current phaseout of HFCs is for 80–85% phaseout by 2035–2045, but there are also plans to halt all development of HFCs before this time. As such, the proposed PQ limit for emissions of Montreal gases ≈0 ODPkg. This can be compared to the sum of ODP kilograms from all Montreal gases emitted for any scale of human activity—the "Montreal gas footprint".

There are two other Planetary Boundary limits that are affected by Montreal Gas emissions:

- Radiative forcing $\leq \pm 1$ Wm^2.
- Extinction rate ≤ 10 extinctions per million species per year.

The limit of zero means that the impacts of Montreal gas on these PBs are eliminated.

11.4   Discussion

It is questionable whether a Planetary Boundary or Planetary Quota is needed for ozone depletion when we are already on the path to recovery. During our extended peer community engagement, it became apparent that not all of the authors of the PBs agreed with its inclusion in that framework.

Ozone depletion is included within the Planetary Quotas for three reasons.

1. There is still a very large hole in the atmosphere that only continued and careful management will resolve.
2. The fundamental scientific basis of the Planetary Quotas is the Planetary Boundaries. Our framework is based on the assumption that this is the best available framework on which to balance this work—we have not undertaken any scientific assessment as to the validity or completeness of the specific PBs. It would thus be incongruent with our project approach to exclude ozone depletion.
3. The purpose of the Planetary Accounting Framework (and thus the Quotas) is to provide a useful mechanism to promote change. The science of behaviour change suggests that humans need and want to see stories of success—that these can motivate humans into action (see Chap. 2). One of the greatest challenges to managing global problems is that people feel the problem is simply too big. The success story of the hole in the ozone and the Montreal Protocol is likely to be a useful tool in helping to generate confidence and action for change. Evidence for how families in regional Western Australia contributed to the global discussions on the ozone hole due to rapidly increasing sunburn with their children is outlined in Stocker et al. (2012).

The journey to living within this PQ started several decades ago. However, we are not at the end of the path. Table 11.2 gives both examples of past actions and potential future actions that have or could occur to help humanity live within this PQ.

Table 11.2 Examples of different scales of activity which have or could contribute to achieving the PQ for Montreal gas emissions

Achieving the planetary quota for Montreal gas emissions

	Community	Government	Business
Large	Develop global treaties to coordinate efforts to phase out ozone depleting problems: e.g., *the Montreal Protocol*	Join global initiatives to collaborate against global environmental problems: e.g., *the Montreal Protocol*	Innovate to provide solutions that allow communities to transition away from harmful activities e.g., *Alternatives to CFCs were developed in light of the Montreal Protocol*
Medium		Ban the sale of ozone-depleting substances e.g., *most countries have now banned most ODPs*	
Small	Lobby for global initiatives against environmental problems e.g., *Australian communities lobbied for ozone protection laws because of high rates of sunburn in children prior to the Montreal Protocol*	Run local campaigns to educate the community about ozone-depleting substances	
Individual	Don't purchase any product with ozone-depleting substances	Propose how to widen the ban on ozone-depleting substances	Ban all ozone-depleting substances from the business

11.5 Conclusions

Ozone-depleting substances are already being phased out under the Montreal Protocol. This phaseout is considered sufficient to meet the Planetary Boundary for ozone depletion.

The Planetary Quota for emissions of OPDs is *emissions of Montreal gasses* ≈ 0 ODPkg. This limit can be compared to the sum of ODPkg of Montreal gasses emitted during any scale of human activity. Humanity are on track to resolving this global environmental problem collectively—in a poly-scalar way (see Chap. 2). This can be used as a success story to motivate action towards living within the other PBs.

References

Chesick JP (1975) Atmospheric halocarbons and stratospheric ozone. Nature 254:275

Chipperfield MP, Fioletov VE, Bregman B, Burrows JP, Connor BJ, Haigh JD, Harris NRP, Hauchecorne A, Hood LL, Kawa SR, Krzyscin JW, Logan JA, Muthama NJ, Polvani L, Randel WJ, Sasaki T, Stähelin J, Stolarski RS, Thomason LW, Zawodny JM (2006) Global ozone: past and present. WMO (World Meteorological Organization). Scientific assessment of ozone depletion. National Oceanic and Atmospheric Administration, National Aeronautics and Space Administration, United Nations Environment Programme. World Meteorological Organization and European Commission, Geneva

Clerbaux C, Cunnold DM, Anderson J, Engel A, Fraser PJ, Mahieu E, Manning A, Miller J, Montzka S, Nassar R, Prinn R, Reimann S, Rinsland CP, Simmonds P, Verdonik D, Weiss R, Wuebbles DA, Yokouchi Y (2006) Long-lived compounds. WMO (World Meteorological Organization). Scientific assessment of ozone depletion. National Oceanic and Atmospheric Administration, National Aeronautics and Space Administration, United Nations Environment Programme, World Meteorological Organization, and European Commission, Geneva

Fahey DW (2003) Twenty questions and answers about the ozone layer. Scientific assessment of ozone depletion: 2002. Global ozone research and monitoring project—report no. 47. World Meteorological Organisation, Geneva

Farman JC, Gardiner BG, Shanklin JD (1985) Large losses of total ozone in Antarctica reveal seasonal ClO_x/NO_x interaction. Nature 315:207–210

Rockström J, Steffen W, Noone K, Persson A, Chapin FS, Lambin E, Lenton TM, Scheffer M, Folke C, Schellnhuber HJ, Nykvist B, De Wit CA, Hughes T, Van Der Leeuw S, Rodhe H, Sorlin S, Snyder PK, Costanza R, Svedin U, Falkenmark M, Karlberg L, Corell RW, Fabry VJ, Hansen J, Walker B, Liverman D, Richardson K, Crutzen P, Foley J (2009) Planetary boundaries: exploring the safe operating space for humanity. Ecol Soc 14:32

Solomon S, Ivy DJ, Kinnison D, Mills MJ, Neely RR, Schmidt A (2016) Emergence of healing in the Antarctic ozone layer. Science 353:269–274

Stocker L, Newman P, Duggie J (2012) Climate change and perth (South West Australia). In: Blakely E, Carbonell A (eds) Resilient coastal regions: planning for climate change in the United States and Australia. Lincoln Institute of Land Policy, Washington, DC

Strahan SE, Douglass AR (2018) Decline in Antarctic ozone depletion and lower stratospheric chlorine determined from aura microwave limb sounder observations. Geophys Res Lett 45:382–390

UNEP (1999) UNEP report on bromide phase out—a handbook for national ozone units

UNEP (2017a) The Montreal Protocol on substances that deplete the ozone layer. http://ozone.unep.org/en/treaties-and-decisions/montreal-protocol-substances-deplete-ozone-layer. Accessed 22 Dec 2017

UNEP (2017b) Treaties and decisions. http://ozone.unep.org/en/treaties-and-decisions. Accessed 22 Dec 2017

Unknown (2015) Earth-atmosphere-layers.jpg. Wikimedia

Chapter 12
A Quota for Aerosols

Air pollution is turning Mother Nature prematurely grey
Irv Kupcinet

Abstract Aerosols are small particles suspended in the atmosphere. They can absorb and scatter light and change cloud formations. They have both warming and cooling impacts, but, overall, the impacts are cooling. They dampen the warming impacts of fossil fuel emissions. However, they can be very harmful to human health.

Until now there has not been an indicator that could link human activity to the abundance of aerosols in the atmosphere. This is because the pathways from the emission of aerosols and precursor gases to aerosols vary greatly and are influenced by several environmental factors such as temperature, humidity, and air movement. However, without a way to even approximate this relationship, it is difficult to effectively manage or limit the source of the emissions.

A new indicator is thus proposed to link the emission of aerosols and precursor gases to aerosol abundance. It is a measure of the equivalent aerosol abundance if emissions occurred at a global scale, in the unit *aerosol optical depth equivalent*.

The new indicator is not intended to estimate the local state of the environment after emissions. Rather, the intent is that the emissions related to an activity can be compared to another activity and to scientific limits at local and global scales.

The Planetary Quota for aerosols is *aerosol optical depth equivalent* between 0.04 and 0.1. This can be compared to the "aerosol footprint" of any scale of human activity. The limit is set on the basis of balancing the need to retain some cooling effects to offset global warming, as well as the need for clean air for the health of humans and other species.

12.1   Introduction

The term *aerosols* describes small particles suspended in the air. Aerosols can be emitted directly (sea salt in the atmosphere is one of the most common naturally occurring aerosols), or they can develop from the emission of precursor gases. The main anthropogenic sources of aerosols are dust (due to desertification) and emissions of sulphur oxides, nitrogen oxides, dimethyl sulphide, organic carbon, black carbon, and volatile organic compounds (Boucher et al. 2013). Atmospheric aerosol loading was included as a Planetary Boundary (PB) because of the influence of aerosols on the climate system through changes to radiative forcing (predominantly cooling) and impacts on human health through air pollution (Rockström et al. 2009). Air pollution has been identified by the World Health Organization (WHO) as the single greatest risk for global health (WHO 2016).

The critical pressures considered within this Quota include emissions of:

- Carbon monoxide
- Non-methane volatile organic compounds
- Nitrates
- Sulphates
- Black carbon
- Organic carbon

These pressures influence not only the Planetary Boundary for aerosol optical depth but also the Boundaries for extinction rate and radiative forcing, i.e. climate change. The latter poses an interesting problem. To return to and remain within the Planetary Boundaries for aerosols and extinction rate, the emissions of aerosols need to be limited as much as possible. However, aerosols have a net cooling effect on the atmosphere. Without the masking effect of aerosols, the earth would be significantly warmer today. As we reduce aerosol pollution, we will need to also reduce greenhouse gas emissions to avoid additional warming impacts. This means that aerosol reduction pathways need to be carefully balanced with emissions reductions pathways.

This chapter begins with a background about aerosols and how these are currently measured. There is no existing indicator which can be used to measure the collective impacts of aerosol and precursor gas emissions. As such, the chapter goes on to introduce the new indicator developed for this purpose—the *aerosol optical depth equivalent*. This is followed by the scientific basis for the proposed Planetary Quota for aerosols. The chapter concludes with a discussion of this PQ in the context of the status quo, including some examples of activities that could help us to live within this PQ.

12.2 Background

Atmospheric aerosols are suspended solid or liquid particles in the air with diameters ranging from a few nanometres to a few tens of micrometres. Aerosols absorb and scatter solar radiation and affect cloud formation. They have both warming and cooling effects on the global climate, but the net effect is cooling.

The idea that aerosols could change global climate dynamics was brought to light in the 1940s when some scientists were concerned that their presence might fast-track the Earth system into another Ice Age (e.g. Rasool and Schneider 1971; Bryson 2009) (see Chap. 2). As scientific understanding of the relative impacts of greenhouse gases and aerosols advanced, it became apparent that the warming impacts of fossil fuels would substantially outweigh the cooling impacts of aerosols. However, the cooling effects of aerosols have substantially masked the warming effects of greenhouse gases; without aerosols the world would be substantially warmer (Boucher et al. 2013).

Aerosols influence the climate system in a complex way. They both scatter and absorb radiation (considered the "direct effects") and modify amounts and properties (microphysical and radiative) of clouds (the "indirect effects") (Chin 2009). Aerosols can both warm and cool Earth's surface, but on average they provide a cooling effect (Boucher et al. 2013). They can be visible as dust, smoke, and haze but can also be invisible to the human eye. Most aerosols come from natural sources such as sea salt and dust. However, human activity is increasing the concentration of aerosols in the atmosphere through the direct emission of aerosols, the emission of precursor gases that result in aerosol formation, and through land use that results in desertification and therefore increased atmospheric dust. Concerns over aerosols in the atmosphere are distinct from past concerns regarding the use of aerosol cans which is an ozone rather than an aerosol problem (see Box 12.1).

Box 12.1 Aerosol Cans

There is common confusion or misconception concerning the environmental impacts of aerosol spray cans.

Aerosol spray cans are named as such because they use high pressure to emit small droplets (aerosols) of liquid. Unlike the aerosols of concern in this section, these aerosols from spray cans do not normally stay suspended in the atmosphere but rather fall to the surface over which they were sprayed. The environmental impacts of aerosol cans are not (usually) related to the aerosols produced but rather due to the compressed gases used to propel the aerosols.

Prior to the Montreal Protocol, aerosol spray cans often used ozone-depleting substances (see Chap. 11) as the propellants. There were widespread campaigns against the use of these cans because of to their contribution to the hole in the ozone layer. Now, most of the ozone-depleting gases previously used in these cans have been phased out and most now use alternative gases that do not contribute to depletion of the ozone layer.

Table 12.1 World Health
Organization guidelines for
ambient air quality

	$PM_{2.5}$	PM_{10}
Annual mean	10 μg/m^3	20 μg/m^3
24-h mean	25 μg/m^3	50 μg/m^3

Many aerosols affect human health including nitrous oxides, ozone, carbon monoxide, and sulphur dioxides. Air pollution has been identified by the World Health
Organization as the single greatest risk for global health (WHO 2016). 5% of all
deaths in 2012 were solely attributable to air pollution (WHO 2016). In 2016, 92%
of the world's population lived in areas that are outside the World Health Organization
ambient air quality recommendations (WHO 2016).

Black carbon is a particularly problematic aerosol due to both its greenhouse gas
impacts and its impacts on human health. Ships in port, planes taxiing before flying,
and trucks travelling slowly through cities are all major producers (Creutzig et al. 2014).

12.3   Measuring Aerosols

The concentration of aerosols in the atmosphere can be quantified using an optical
measure (i.e., the amount of light which can pass through the atmosphere) or by
mass concentration (i.e., the mass of aerosols per volume of atmosphere):

- Aerosol optical depth (AOD)—also known as aerosol optical thickness is an
  optical measure. It is a dimensionless unit that expresses the fraction of inci-
  dent light either scattered or absorbed by airborne particles in a vertical column of air (Chin 2009). An AOD value of zero indicates completely clear
  skies. An AOD of one indicates that no light can permeate the atmosphere
  (Chin 2009). The Planetary Boundary for aerosol loading is a maximum
  regional AOD of 0.25—with an increase due to human activity ≤0.1 (Steffen
  et al. 2015). The global mean value at 550 nm is approximately 0.12–0.16
  (Chin et al. 2014).
- Particulate matter concentration (PMC) is a mass concentration measure of
  aerosols. It is a measure of the number of grams of particulate per volume of
  air (μg/m^3). PMC is often reported for particulate matter of a specific size,
  commonly with diameters less than 2.5 μm and 10 μm ($PM_{2.5}$ and PM_{10}). $PM_{2.5}$
  is the most harmful category of aerosol with respect to human health (Fantke
  et al. 2015). For this reason, it is often used as a proxy indicator for air pollution related to health (WHO 2016).

The World Health Organization uses the metrics $PM_{2.5}$ and PM_{10} to communicate
guidelines for minimum air quality standards (see Table 12.1). These guidelines are
for maximum particulate concentration. The WHO position is that there is no level
of particulate matter that does not have any impacts on human health. They therefore recommend that target levels are as low as possible (WHO 2016).

These units both pertain to the *state* of the environment. Until now, there has not been an indicator with which to collectively measure aerosols at a *pressure* level in a way that can be scaled.[1]

Fantke et al. (2015) identified the importance of assessing $PM_{2.5}$ health impacts in environmental impact assessments. They chose $PM_{2.5}$ because this has the most severe impacts on human health (Harrison and Yin 2000; Lim et al. 2012; Lippmann and Chen 2009). They developed a framework to include the health impacts of $PM_{2.5}$ into life-cycle assessment. Their framework considers the amount of primary and secondary particulate matter that is taken in by people. An exposure response factor and a severity factor are then applied to determine a human health-related impact score in the health unit *disability-adjusted life years* (DALY).

Fantke et al.'s (2015) framework is extremely useful. It provides a quantitative measure of human health impacts from various activities. However, as is the problem with any life-cycle assessment indicator, there is no clear limit. How many DALYs are acceptable for a given product? It is also specific to the location of the emissions. The intake fraction of $PM_{2.5}$ will vary greatly depending on population density, proximity to the activity, and local climate (Humbert et al. 2011). Poor air quality is a local problem. It is also a global problem. Aerosols in the atmosphere are affecting the climate system. Air quality affects humanity directly through health impacts and indirectly through impacts on Earth-system functioning. Fantke et al.'s framework has strong local relevance. It was developed with a focus on health, not on global Earth-system impacts. It is not straightforward to relate it to the Planetary Boundary for aerosols. Nor could it easily be scaled and adapted for use in a poly-scalar approach.

Gronlund et al. (2015) also look at assessing the human impacts of $PM_{2.5}$ through a characterization factor—impact per kg of $PM_{2.5}$ emitted. Their approach has a similar health approach and thus similar limitations to Fantke et al.'s measure.

In a study linking life-cycle impact assessment to the Planetary Boundaries, Ryberg et al. (2018) proposed characterization factors (CFs) to link key environmental flows to Planetary Boundaries. Their work includes characterization factors linking emissions of key aerosols and precursor gases to AOD, in kilograms per year. The CFs are estimated at global and regional levels and express a change in AOD per annual mass of aerosol emissions. These CFs thus link the pressure of aerosol and precursor emissions at a regional scale to the *state* of aerosols in the atmosphere (aerosol optical depth).

---

[1]A *pressure level* means at the level of environmental flows, i.e. the emission of aerosols and precursor gases. See Chap. 4 for a description of different categories of environmental indicators including states and pressures.

12.4   Equivalent Aerosol Optical Depth

The pathways from the emission of an aerosol or a precursor gas are complex. The pathways vary with local environmental conditions. Aerosols and precursor gases also interact with one another. This makes it prohibitively difficult to accurately estimate the impacts of the emissions of a given substance to local AOD levels or to PMC without complex computer modelling.

The CFs proposed by Ryberg et al. (2018) are based on very simplified calculations. The atmospheric transport of aerosols is perhaps not adequately captured. However, the complexity of interactions between different aerosols and the lifetimes of different aerosols and precursors are considered, albeit simplified. It would be highly inaccurate to suggest that using these calculations one could predict the resulting AOD.

However, the framework has substantial merit in that it links the *pressure* of aerosol and precursor emissions to a *state*. Building on their approach, it is possible to estimate the contribution of an activity to global average AOD. This should not be confused with an estimation of actual change in AOD. Such an estimation would be highly inaccurate because of variations to local conditions and the interactions between different aerosols and precursors.

By *equating emissions to impacts on the global average, one can effectively estimate the equivalent impacts of emissions, or **equivalent AOD (AODe)**.*

This concept is not dissimilar to the way we currently measure greenhouse gases or ozone-depleting substances. Greenhouse gas emissions can be assessed for their global warming impacts by equating them to an amount of carbon dioxide (CO_2) that would warm the atmosphere by the same amount—the *equivalent* CO_2 (CO_2e). In the same vein, the ozone-depleting potential (ODP) of ozone-depleting substances is measured with respect to a benchmark gas. Emitting 1 kg of nitrous oxide, with an ODP of 0.17, is equivalent to emitting 0.17 kg of CFC-11—the benchmark gas. The premise of AODe differs slightly from the examples given in that the equivalency is not set against a substance but against an effect, i.e. change in AOD. However, the basis of the unit is of similar origin to CO_2e and ODP.

12.4.1   Calculating AODe

AOD can be calculated using the formula:

$$AOD = MEE \times M \tag{12.1}$$

MEE is the mass extinction efficiency or specific extinction in m^2/g and M is the aerosol mass loading per unit surface area in g/m^2.

The CFs developed by Ryberg et al. (2018) are based on the derivation of aerosol mass loading for a given activity multiplied by the specific extinction (at a certain relative humidity) derived from Chin et al. (2002). Mass loading for a given substance (*n*) is estimated using Eq. (12.2), where E denotes average emissions in kg/year, τ denotes residence time in years, and A denotes global (or regional) terrestrial area in m^2.

$$M_n = E_n \times \frac{\tau_n}{A} \qquad (12.2)$$

The CF is then given by Eq. (12.3) where β denotes the specific extinction efficiency.

$$CF_n = \beta_n \times M_n \qquad (12.3)$$

Building on this approach, Ryberg and Meyer have derived an alternative method to estimate AODe.[2] The mass loading for AODe is calculated using Eq. (12.4), where $A(x)$ represents corresponding area. For example, if estimating the AODe for an individual, $A(x)$ could be a per capita share of global terrestrial area.[3]

$$M_n = E_n \times \frac{\tau_n}{A_x} \qquad (12.4)$$

The resultant AODe for substance n is then determined using Eq. (12.5) and total AODe from an annual emission flux using Eq. (12.6).

$$AODe_n = \beta_n \times M_n \qquad (12.5)$$

$$AODe_{steady\ state} = \Sigma\, AODe_n \qquad (12.6)$$

12.5   The Limit

There are three Planetary Boundaries that correspond with this Planetary Quota:

- Aerosol optical depth ≤ 0.1
- Radiative forcing $\leq \pm 1$ Wm^2
- Extinction rate ≤ 10 extinctions per million species per year

There is no global PB limit defined for aerosols; however, in the most recent update of the boundaries, a regional limit of aerosol optical depth (AOD) ≤ 0.25 was proposed. To account for the fact that many aerosols occur naturally, a

---

[2] Manuscript in preparation.
[3] Determining the appropriate area will depend on the allocation procedure selected for downscaling the global quotas. See Chap. 17 for more on allocation procedures.

specific limit for anthropogenic aerosols was also defined: $AOD_{anthro} \leq 0.1$ (Steffen et al. 2015). This limit was set on the basis of limiting impacts on the ocean-atmospheric circulation (Steffen et al. 2015).

As discussed in Chap. 8, it is not possible to derive a specific limit for radiative forcing for different forcing elements from the PB for radiative forcing. However, based on the Planetary Quotas determined for carbon dioxide, methane and nitrous oxide (Ag-GHGs), forestland, and ozone-depleting substances in Meyer and Newman (2018) and Meyer (2018), a range of acceptable radiative forcing levels from the PQ for aerosols can be determined.

There are also no specific air quality guidelines pertaining to species extinctions. However, as discussed in Sect. 12.3, $PM_{2.5}$ is often used as a proxy for air quality. The WHO guideline for human health is $PM_{2.5} \leq 10 \ \mu g/m^3$. This limit is assumed to be an acceptable proxy limit for other species.

There have been studies linking AODe to both radiative forcing (e.g. Hansen et al. 2005; Andersson et al. 2015) and to $PM_{2.5}$ (e.g. Engel-Cox et al. 2004; Liu et al. 2004; Gupta and Christopher 2009; Gupta et al. 2006, 2013; van Donkelaar et al. 2010).

12.5.1   Radiative Forcing

The radiative forcing from the PQs for carbon dioxide, Ag-GHGs, forestland (Chap. 10), and ozone-depleting substances (Chap. 11) are shown below.

12.5.1.1   Carbon Dioxide and Ag-GHGs

The combined forcings estimated for the original Quotas for CO_2 and Ag-GHG is 1.73 W/m^2 (Meyer and Newman 2018; Meyer 2018). There is limited data to connect the new limits proposed in this book to radiative forcing. However, it is assumed here that the amendments to the Quotas for carbon dioxide and Ag-GHGs have a negligible net effect on long-term radiative forcing impacts.

12.5.1.2   Other Greenhouse Gases

HFCs, PFCs, and SF_6 were excluded from the list of critical pressures because they currently each contribute less than 1% towards total radiative forcing. However, in the future, their relative contribution could be much higher. Indicative radiative forcing values for these are based on RCP2.6 projections for 2100 to give 0.142 W/m^2 (HFCs—0.126 W/m^2 and PFCs and SF_6 combined of 0.016 W/m^2) (IPCC 2013).

12.5.1.3   Forestland

It is not straightforward to predict the future albedo (surface reflectivity) of the Earth. To meet the PB for land use would require approximately one billion hectares of reforestation. However, it is difficult to estimate the areas of ice, albedo of future urban areas, total future cropland areas, etc. The change in land use since 1870 led to a change in albedo with a radiative forcing impact estimated at -0.15 ± 0.1 W/m^2 (Myhre et al. 2013). Major reforestation would reduce the albedo and therefore have a positive forcing effect. To determine a rough approximation for future albedo forcing, it is assumed that this will be of a similar order of magnitude as changes since 1870, but in the opposite direction. Thus, the estimated radiative forcing based on the increase in forestland is approximately 0.15 W/m^2.

12.5.1.4   Ozone

The PQ for ozone is zero. As such, the forcing is also zero.

Combining the estimated forcings above gives a total (excluding aerosol impacts) of 2.25 W/m^2. This means that to respect the PB for radiative forcing of $\leq \pm 1$ W/m^2, the radiative forcing impacts of aerosols would need to be ≤ -1.25 W/m^2 (to a minimum of -3.25).

Radiative forcing due to stratospheric aerosols depends predominantly on the aerosol optical depth. The adjusted forcing due to aerosols can be approximated using Eq. (12.7) (Andersson et al. 2015; Sato et al. 1993; Hansen et al. 2005).

$$-25 \times AOD \approx RF_{Aero} \tag{12.7}$$

Thus, to respect the PB for radiative forcing, the PQ for aerosols must be $0.04 \leq AODe \leq 0.13$.

12.5.2   Air Pollution

There have been several studies looking at the relationship between AOD and PM$_{2.5}$ including Engel-Cox et al. (2004), van Donkelaar et al. (2010), Gupta et al. (2006, 2013) and Liu et al. (2004). The simplest relationships are given by a two-variable regression equation. AOD values obtained using the WHO PM$_{2.5}$ recommendation for annual concentration limits of 10 μg in a sample of two-variable regressions listed in Gupta et al. (2013) give results as shown in Table 12.2.

The highest value in this range is based on the relationship proposed by Gupta et al. (2006). The same author later derived an alternative formula (see Table 12.2, line 2) which gives much more congruent results with the other equations. This later proposal by Gupta suggests a change or advance in thinking; as such, his earlier proposal can be discounted.

Table 12.2 AOD values according to various two-variable regression equations

Formula	AOD	References
$PM_{2.5} = 7.54 + 18.66AOD$	0.14	Engel-Cox et al. (2004)
$PM_{2.5} = 87.5AOD$	0.114	Derived by Gupta et al. (2013) from van Donkelaar et al. (2010)
$AOD = 0.006 \times PM_{2.5} + 0.149$	0.209	Gupta et al. (2006)
$PM_{2.5} = 81AOD$	0.123	Liu et al. (2004)

All of the calculations indicate that a limit based directly on the PB limit (i.e., AODe $\leq$ 0.1) would respect the WHO air quality recommendations. This is consistent with the academic literature which typically refers to AOD values of this order of magnitude as low or pertaining to clear skies (e.g. Gupta et al. 2013; Engel-Cox et al. 2004; NOAA n.d.).

12.5.3   The PQ for Aerosols

Considered in isolation, the lower the AODe the better. However, as aerosols provide a predominantly cooling forcing, they also offset some of the warming impacts of GHG emissions and land-use change. It has been estimated that aerosols reduce solar energy at the ground by as much as 8% in densely populated areas (Alpert and Kishcha 2008). Without aerosols, average global temperatures would be higher (Boucher et al. 2013).

Thus, the PQ for aerosols comprises both a minimum to offset radiative forcing impacts and a maximum to limit impacts on human health. This gives a PQ for aerosols of 0.04 $\leq$ AODe $\leq$ 0.1. This can be compared with the "aerosol footprint" of any scale of human activity—i.e. the annual AODe associated with the activity.

The argument could be made that further GHG reductions and increased reforestation would allow the PQ for aerosols to be lowered further. However, the current PQs for GHGs and reforestation are extremely ambitious. Given that the PB for aerosols and the WHO health guidelines can be met with the proposed PQ for aerosols, it does not seem worthwhile to push the PQs for GHGs and reforestation further at this stage.

12.6   Discussion

There is no data on current AODe. Estimates of global mean AOD values of 0.12–0.16 (Chin et al. 2014) do not distinguish human-induced aerosols from naturally occurring aerosols. However, given these global mean values, we can deduce that we have not exceeded the PQ for aerosols at a global scale. Regional AODe is likely to be above the PQ level for many industrial and/or highly populated locations.

Table 12.3 Examples of different scales of activity which have or could contribute to achieving the PQ for aerosols

Achieving the Planetary Quota for aerosol and precursor emissions			
	Community	**Government**	**Business**
Large	Develop guidelines for minimum healthy air quality, e.g. *the WHO minimum standards for clean air*	Set minimum air quality laws and implement initiatives to manage polluters, e.g. *German laws on air quality control limit emissions of relevant air pollutants in new installations and require that existing installations must be upgraded*	Improve business practices to reduce and eventually eliminate the emission of aerosols and precursor gases
Medium	Install a community renewable energy plant	Manage local air quality, e.g. *Stuttgart local government reduces public transport fares to half rates when air pollution levels exceed a certain level to encourage citizens out of cars*	Use electric vehicles to transport planes to eliminate emissions from taxiing
Small	Transition to fossil fuel free power	Educate communities about how to make choices for better air quality	Upgrade truck fleets
Individual	Walk or train instead of driving	Behaviour change programs	Develop technology to reduce reliance on aerosol and precursor gas emitting products and services

Table 12.3 lists examples of activities for different scales of activity across different sectors which either have already or could in future contribute towards managing human activity within the PQ for aerosols.

12.7 Conclusions

Aerosols suspended in the atmosphere are harmful to human health and are affecting the global climate. There was not previously an indicator that collectively measured the emissions of aerosols and precursors in a way that could be applied to different scales of activity.

This chapter introduced the new metric AODe, a measure of the relative impacts of aerosols and precursors on the atmospheric aerosol depth. The Planetary Quota for aerosols is $0.04 \leq AODe \leq 0.1$. The upper limit is set to minimize impacts of aerosols on human health. The lower limit is to continue to offset warming impacts from other forcing agents.

References

Alpert P, Kishcha P (2008) Quantification of the effect of urbanization on solar dimming. Geophys Res Lett 35:L08801

Andersson SM, Martinsson BG, Vernier J-P, Friberg J, Brenninkmeijer CAM, Hermann M, Van Velthoven PFJ, Zahn A (2015) Significant radiative impact of volcanic aerosol in the lowermost stratosphere. Nat Commun 6:7692

Boucher O, Randall D, Artaxo P, Bretherton C, Feingold G, Forster P, Kerminen VM, Kondo Y, Liao H, Lohmann U, Rasch P, Satheesh SK, Sherwood S, B S, Zhang XY (2013) Clouds and aerosols. In: Stocker TF, Qin D, Plattner G-K, Tignor M, Allen SK, Boschung J, Nauels A, X Y, Bex V, Midgley PM (eds) Climate change 2013: the physical science basis. Contribution of working group I to the fifth assessment report of the intergovernmental panel on climate change. Cambridge University Press, Cambridge

Bryson RA (2009) The lessons of climatic history. Environ Conserv 2:163–170

Chin M (2009) Atmospheric aerosol properties and climate impacts. Diane Publishing, Collingdale, PA

Chin M, Ginoux P, Kinne S, Torres O, Holben B, Duncan B, Martin R, Logan J, Higurashi A, Nakajima T (2002) Tropospheric aerosol optical thickness from the GOCART model and comparisons with satellite and sun photometer measurements. J Atmos Sci 59:461–483

Chin M, Diehl T, Tan Q, Prospero JM, Kahn RA, Remer LA, Yu H, Sayer AM, Bian H, Geogdzhayev IV, Holben BN, Howell SG, Huebert BJ, Hsu NC, Kim D, Kucsera TL, Levy RC, Mishchenko MI, Pan X, Quinn PK, Schuster GL, Streets DG, Strode SA, Torres O, Zhao XP (2014) Multi-decadal aerosol variations from 1980 to 2009: a perspective from observations and a global model. Atmos Chem Phys 14:3657–3690

Creutzig F, Cruz-Núñez X, D'agosto M, Dimitriu D, Figueroa Meza MJ, Fulton L, Kobayashi S, Lah O, Mckinnon A, Newman P, Ouyang M, Schauer JJ, Sperling D, Tiwari G (2014) Transport. In: Edenhofer O, Pichs-Madruga R, Sokona Y, Farahani E, Kadner S, Seyboth K, Adler A, Baum I, Brunner S, Eickemeier P, Kriemann B, Savolainen J, Schlömer S, Von Stechow C, Zwickel T, Minx JC (eds) Climate change 2014: mitigation of climate change. Contribution of working group III to the fifth assessment report of the intergovernmental panel on climate change. Cambridge University Press, Cambridge

Engel-Cox JA, Holloman CH, Coutant BW, Hoff RM (2004) Qualitative and quantitative evaluation of Modis satellite sensor data for regional and urban scale air quality. Atmos Environ 38:2495–2509

Fantke P, Jolliet O, Evans JS, Apte JS, Cohen AJ, Hänninen OO, Hurley F, Jantunen MJ, Jerrett M, Levy JI, Loh MM, Marshall JD, Miller BG, Preiss P, Spadaro JV, Tainio M, Tuomisto JT, Weschler CJ, Mckone TE (2015) Health effects of fine particulate matter in life cycle impact assessment: findings from the Basel Guidance Workshop. Int J Life Cycle Assess 20:276–288

Gronlund C, Humbert S, Shaked S, O'Neill M, Jolliet O (2015) Characterizing the burden of disease of particulate matter for life cycle impact assessment. Int J 8:29–46

Gupta P, Christopher SA (2009) Particulate matter air quality assessment using integrated surface, satellite, and meteorological products: multiple regression approach. J Geophys Res 114:D14205. https://doi.org/10.1029/2008JD011496

Gupta P, Christopher SA, Wang J, Gehrig R, Lee Y, Kumar N (2006) Satellite remote sensing of particulate matter and air quality assessment over global cities. Atmos Environ 40:5880–5892

Gupta P, Khan MN, da Silva A, Patadia F (2013) Modis aerosol optical depth observations over urban areas in Pakistan: quantity and quality of the data for air quality monitoring. Atmos Pollut Res 4:43–52

Hansen J, Sato M, Ruedy R, Nazarenko L, Lacis A, Schmidt GA, Russell G, Aleinov I, Bauer M, Bauer S, Bell N, Cairns B, Canuto V, Chandler M, Cheng Y, Del Genio A, Faluvegi G, Fleming E, Friend A, Hall T, Jackman C, Kelley M, Kiang N, Koch D, Lean J, Lerner J, Lo K, Menon S, Miller R, Minnis P, Novakov T, Oinas V, Perlwitz J, Perlwitz J, Rind D, Romanou

A, Shindell D, Stone P, Sun S, Tausnev N, Thresher D, Wielicki B, Wong T, Yao M, Zhang S (2005) Efficacy of climate forcings. J Geophys Res Atmos 110

Harrison RM, Yin J (2000) Particulate matter in the atmosphere: which particle properties are important for its effects on health? Sci Total Environ 249:85–101

Humbert S, Marshall JD, Shaked S, Spadaro JV, Nishioka Y, Preiss P, Mckone TE, Horvath A, Jolliet O (2011) Intake fraction for particulate matter: recommendations for life cycle impact assessment. Environ Sci Technol 45:4808

IPCC (2013) Annex II: climate system scenario tables. In: Stocker TF, Qin D, Plattner G-K, Tignor M, Allen SK, Boschung J, Nauels A, Xia Y, Bex V, Midgley PM (eds) Climate change 2013: the physical science basis. Contribution of working group I to the fifth assessment report of the intergovernmental panel on climate change. Cambridge University Press, Cambridge

Lim SS, Vos T, Flaxman AD, Danaei G, Shibuya K, Adair-Rohani H, Almazroa MA, Amann M, Anderson HR, Andrews KG, Aryee M, Atkinson C, Bacchus LJ, Bahalim AN, Balakrishnan K, Balmes J, Barker-Collo S, Baxter A, Bell ML, Blore JD, Blyth F, Bonner C, Borges G, Bourne R, Boussinesq M, Brauer M, Brooks P, Bruce NG, Brunekreef B, Bryan-Hancock C, Bucello C, Buchbinder R, Bull F, Burnett RT, Byers TE, Calabria B, Carapetis J, Carnahan E, Chafe Z, Charlson F, Chen H, Chen JS, Cheng AT-A, Child JC, Cohen A, Colson KE, Cowie BC, Darby S, Darling S, Davis A, Degenhardt L, Dentener F, Des Jarlais DC, Devries K, Dherani M, Ding EL, Dorsey ER, Driscoll T, Edmond K, Ali SE, Engell RE, Erwin PJ, Fahimi S, Falder G, Farzadfar F, Ferrari A, Finucane MM, Flaxman S, Fowkes FGR, Freedman G, Freeman MK, Gakidou E, Ghosh S, Giovannucci E, Gmel G, Graham K, Grainger R, Grant B, Gunnell D, Gutierrez HR, Hall W, Hoek HW, Hogan A, Hosgood HD, Hoy D, Hu H, Hubbell BJ, Hutchings SJ, Ibeanusi SE, Jacklyn GL, Jasrasaria R, Jonas JB, Kan H, Kanis JA, Kassebaum N, Kawakami N, KhanG Y-H, Khatibzadeh S, Khoo J-P, Kok C (2012) A comparative risk assessment of burden of disease and injury attributable to 67 risk factors and risk factor clusters in 21 regions, 1990–2010: a systematic analysis for the Global Burden of Disease Study 2010. The Lancet 380:2224–2260

Lippmann M, Chen L-C (2009) Health effects of concentrated ambient air particulate matter (CAPs) and its components. Taylor & Francis, Boca Raton

Liu Y, Park RJ, Jacob DJ, Li Q, Kilaru V, Sarnat JA (2004) Mapping annual mean ground-level PM 2.5 concentrations using Multiangle Imaging Spectroradiometer aerosol optical thickness over the contiguous United States. J Geophys Res Atmos 109

Meyer K (2018) Planetary quotas and the planetary accounting framework—comparing human activity to global environmental limits. PhD, Curtin

Meyer K, Newman P (2018) The planetary accounting framework: a novel, quota-based approach to understanding the planetary impacts of any scale of human activity in the context of the Planetary Boundaries. Sustainable Earth 1

Myhre G, Shindell D, Bréon F-M, Collins W, Fuglestvedt J, Huang J, Koch D, Lamarque J-F, Lee D, Mendoza B, Nakajima T, Robock A, Stephens G, Takemura T, Zhang H (2013) Anthropogenic and natural radiative forcing. In: Stocker TF, Qin D, Plattner G-K, Tignor M, Allen SK, Boschung J, Nauels A, Xia Y, Bex V, Midgley PM (eds) Climate change 2013: the physical science basis. Contribution of working group I to the fifth assessment report of the intergovernmental panel on climate change. Cambridge University Press, Cambridge

NOAA (n.d.) Surfrad aerosol optical depth. http://www.esrl.noaa.gov/gmd/grad/surfrad/aod/. Accessed 4 Oct 2016

Rasool SI, Schneider SH (1971) Atmospheric carbon dioxide and aerosols: effects of large increases on global climate. Science 173:138–141

Rockström J, Steffen W, Noone K, Persson A, Chapin FS, Lambin E, Lenton TM, Scheffer M, Folke C, Schellnhuber HJ, Nykvist B, De Wit CA, Hughes T, Van Der Leeuw S, Rodhe H, Sorlin S, Snyder PK, Costanza R, Svedin U, Falkenmark M, Karlberg L, Corell RW, Fabry VJ, Hansen J, Walker B, Liverman D, Richardson K, Crutzen P, Foley J (2009) Planetary boundaries: exploring the safe operating space for humanity—supplementary information. Ecol Soc 14

Ryberg MW, Owsianiak M, Richardson K, Hauschild MZ (2018) Development of a life-cycle impact assessment methodology linked to the Planetary Boundaries framework. Ecol Indic 88:250–262

Sato M, Hansen JE, McCormick MP, Pollack JB (1993) Stratospheric aerosol optical depths, 1850–1990. J Geophys Res Atmos 98:22987–22994

Steffen W, Richardson K, Rockström J, Cornell SE, Fetzer I, Bennett EM, Biggs R, Carpenter SR, De Vries W, De Wit CA, Folke C, Gerten D, Heinke J, Mace GM, Persson LM, Ramanathan V, Reyers B, Sörlin S (2015) Planetary boundaries: guiding human development on a changing planet. Science 347:736

Van Donkelaar A, Martin RV, Brauer M, Kahn R, Levy R, Verduzco C, Villeneuve PJ (2010) Global estimates of ambient fine particulate matter concentrations from satellite-based aerosol optical depth: development and application. Environ Health Perspect 118:847

WHO (2016) Ambient air pollution: a global assessment of exposure and burden of disease. World Health Organisation Press, Geneva

Chapter 13
A Quota for Water

Water is the driving force for all of nature
Leonardo da Vinci

Abstract Water is a unique resource in that it is essential to life and irreplaceable. The water cycle is a critical Earth-system process that human activity is beginning to alter. Water availability varies significantly across the globe. There is an abundance of water in some places, and extreme shortages in others. This regionality has led to some debate as to the existence of a global limit for water.

The regional variability of water scarcity does not mean that water is not a global commodity. Water used directly by a consumer is only a small proportion of her total water use. Water is also used indirectly in the production of goods and services as "virtual water". Approximately 40% of the water consumed in Europe is virtual water. It is not a rational argument to suggest that those in water-rich locations need not be concerned about water consumption as much of the water they consume is likely to be from other locations.

The Planetary Boundary for water is only for blue water, i.e. it excludes the use of green water (rainwater) and grey water (contaminated water). Blue water consumption is a reasonable proxy indicator with which to understand the state of the world's water assets. However, the Planetary Quota for water needs to be in a unit that makes sense across different scales of human activity. As such, the use of green water and production of grey water are both relevant and important. Further, the

We would like to make particular acknowledgement of Professor Arjen Hoekstra's contributions to this chapter. Arjen and author Kate Meyer spent a happy afternoon debating the relative merits of different approaches to a water quota during her extended peer review with the scientific community. Kate and Arjen remained in touch and Arjen became one of the Planetary Accounting Network's Founding Members. Arjen's contributions and insights were very influential in the final proposals for both the metric and the limit for water and into the development of Planetary Accounting overall. Devastatingly Arjen passed away late 2019 during a cycle commute in the Netherlands at only 52 years of age. Arjen made huge contributions to his fields of research and we are very pleased to have had the opportunity to gain his knowledge in the development of Planetary Accounting.

Planetary Boundary for water considers gross water consumption. The level of water treatment now available is such that net water consumption is substantially lower than gross water consumption. It is also more relevant to planetary health.

There is no consensus as to a global water budget for net blue, green, and grey water. However, some argue that even at current consumption rates many of our global water bodies are under stress suggesting that the upper limit cannot be higher than current consumption rates.

Thus, the Planetary Quota for water is net water (blue, green, and grey water) ≤ 8500 km^3. This limit is set based on the current global water footprint and can be compared to the water footprint of any scale of activity.

13.1   Introduction

Water is a unique resource as there are no substitutes for most of its uses (Postel et al. 1996). It is impractical to transport it further than a few hundred kilometres in its virgin form, although the transportation of embodied water in food and products is commonplace. The total amount of water on the planet does not change. However, humans can impact both the accessibility and the quality of water.

The critical pressures pertaining to water include:

- Water consumption
- Use of chemicals
- Disposal of chemicals

This chapter begins with an overview of the water cycle and an explanation of the different categories of water (green, blue, and grey). This is followed by the justification for a global water boundary (as opposed to regional boundaries). The main body of the chapter is dedicated to presenting the case for the water indicator selected and the corresponding limit. The chapter concludes with a discussion about what the PQ for water could mean in practice for society today.

13.2   Background

Freshwater is essential to human survival. Not only because we need to drink it, but also because it is needed to produce food. Approximately 90% of water used by humans is used for agriculture. Inland fisheries are critical sources of nutrition—particularly in landlocked countries. Humans also use water as a source of power, for hygiene, and for recreational purposes.

When viewed from outer space, it seems that Earth is abundant with water. Over 70% of Earth's surface is water. However, 97.5% of the world's water is saline. Of the remaining 2.5%, some 35 million km^3 of freshwater, approximately 24 million km^3 (69%), is frozen (Postel et al. 1996). This leaves approximately 11 million km^3 of freshwater which is located in aquifers, soil pores, lakes, swamps, rivers, plant life, and the atmosphere (Secretariat of the CBD 2001; Shiklomanov 1993).

Water can be divided into renewable and "fossil" water. Renewable water is water that flows through the solar-powered hydrological cycle, i.e. from the atmosphere, to rainwater, then water stored in rivers and lakes and some groundwater aquifers. This water has a mean residence time in each state of approximately 2.5 weeks (Oki and Kanae 2006). The term fossil water refers to water which has been stored underground, undisturbed for millennia. It can be tapped; however, recharge takes hundreds or thousands of years (Postel et al. 1996). Accessing this water is thus depleting reserves and can be likened to our depletion of oil wells.

Salt water can be turned into freshwater through a process called desalination. 0.1% of the world's water supply in 1990 was desalinated water (Wangnick Consulting 1990). The problem with desalination is that it is energy intensive. The theoretical minimum energy needed is just under 1 kWh/m^3 of water (Postel et al. 1996). Current best practice is between 2.5 and 3.5 kWh/m^3 (AMTA 2016).

There are several different ways to talk about water. Freshwater is either referred to as green water or blue water. There are also two other categories of water, grey water, and virtual water.

Green water is precipitation on land which does not run off or recharge groundwater.

Blue water is fresh surface water and groundwater, i.e. the water found in freshwater lakes, rivers and aquifers.

The *grey water footprint* is the amount of water that would be needed to dilute pollutants in water to meet specific water standards (Mekonnen and Hoekstra 2011b). The term *grey water* is also used to describe waste water from sinks and showers (as opposed to waste water from kitchens and toilets which is known as black water). This is not the same as the grey water footprint.

Virtual water is the term used to describe the water that is used in the production and transportation of goods and services but is not actually contained in the final product. For example, a lot of water is needed during the extraction of coal from coal mines. The coal will then be transported to a power plant, which also uses water in the generation of electricity. This water is not delivered as water to homes but was fundamental in producing the electricity. It is thus considered the *virtual water*.

Three key risks have been identified regarding human manipulation of the water cycle (Rockström et al. 2009):

1. Water consumption that alters volumes and flow patterns of water bodies;
2. Loss of soil moisture; and
3. Decline in moisture feedback from vapour flows.

The first risk is about over-appropriation of blue water. The second pertains to the disruption of the green water cycle.

Water bodies decline and are replenished naturally. They can feed and be fed by rivers and streams. Rain can replenish water bodies and evaporation can reduce them. If humans withdraw water at a rate higher than the natural cycle can replenish, water bodies can begin to dry up (see Box 13.1). This might mean habitat loss for aquatic ecosystems, diminished water supply for downstream needs or a threat to water availability for ongoing human consumption.

Loss of soil moisture occurs through land-use change. Tree and shrub roots allow soil to hold water and release it through evapotranspiration. Grass, or sandy surfaces, cannot retain water for long and quickly release the water to groundwater aquifers. Plants and plant litter reduce the rate of evaporation of soil water. Deforestation, or land-use which leads to degradation of the land, can thus alter the amount of water the soil can hold. Less moisture in the soil can limit plant growth and therefore carbon uptake.

Moisture feedback from the land and water bodies back to the atmosphere is an important part of climate regulation. A decline in moisture feedback (evaporation and evapotranspiration) can lead to changes in local and regional rainfall patterns.

13.3   A Global Problem

Water availability varies from region to region. In some areas, such as Southern Africa and Australia, water is scarce and droughts common (see Box 13.1). In other areas, local water availability is plentiful.

Box 13.1 Day Zero

Cape Town was predicting that the municipal water supply would shut down and taps run dry in 2019. The day this would happen was called "Day Zero".

There have been predictions that Cape Town could run out of water since 1990. The cause of the water shortage is thought to be a combination of population growth (and therefore increased demand) and a drought in the Western Cape of South Africa which started in 2015 and is thought to be an impact of climate change similar to reduction in rainfall in Perth (similarly located at the end of the continent with the "roaring forties" moving south).

Current water consumption in Cape Town is approximately 200 billion litres per year (Pitt 2018). The goal was to reduce this to 165 billion—a ration of approximately 50 L of water per person per day (Pitt 2018). The World Health Organization suggests that 50–100 L are needed per person per day to ensure that most basic needs are met (UN 2015).

Despite the water crisis, water is still being exported from the region as virtual water. In 2016, 428 billion litres were used in the production of wine for export, and 112 billion litres were used for citrus exports (Leahy 2018).

Cape Town avoided the Day Zero, but only just, by drastically reducing consumption before the rains arrived. Perth similarly avoided its apocalyptic end due to collapse from lack of water by building desalination systems powered by wind farms. These now make 60% of Perth's water from the Indian Ocean. The crisis is not over in either city and the next phase in Perth is to recycle sewage into groundwater supplies and to drastically reduce water consumption through urban and building design and behaviour programs.

The regional availability of water has led to much controversy over the existence of a global limit for water. The main argument for those who do not believe in a global limit is that it does not make sense for those with abundant water supplies to limit their showers and irrigation when the scale of their water consumption has negligible impact on the water bodies they are sourcing this from. They argue that there is no feasible way for water-rich countries to transport their water to water-scarce countries, and as such, it does not make sense to consider this issue at a global scale. They reason that water saving measures should be prioritized in water-stressed areas (Ridoutt and Huang 2012).

Those who argue that a global limit does exist do not disagree that water savings should be prioritized in water-stressed areas. However, they argue against the premise that water is not transportable. They contend that we transport virtual water, water used in goods and services, all over the world (see Fig. 13.1). They reason that it is common for the use of products in locations far removed from the point of virtual water consumption and that water is thus a global resource.

The Planetary Boundary (and the Planetary Quota) for water are based on the second point of view described above, i.e. that water is a global resource. The need for different responses in different situations (i.e. the need to urgently address water consumption in water-scarce regions) is true for all of the PB to a varying extent. Countries with high reliance on fossil fuel energy will need to take greater and more urgent action towards reducing emissions than those in countries with mostly renewable energy, for example. The argument of regional variability is present for any PB or PQ. The purpose of the PBs is to identify which Earth-system processes humans are altering that could put us at risk of changing the state of the Earth system. The water cycle is one of these processes.

Moreover, there are substantial efficiencies to be made by considering water as a global resource. This is likely to be one of the mechanisms which will help to resolve the Planetary Quota over-appropriation of water in water-scarce regions. Governments typically look at water from a national perspective rather than considering impacts or opportunities from virtual water imports (Mekonnen and Hoekstra 2011b). Yet, agricultural trade saved global water consumption of approximately 369 teralitres per year between 1996 and 2005.

Of the water saved through agricultural virtual water, approximately 59% was green water, 27% blue and 15% grey. The global blue water savings achieved account for 10% of the total global blue water footprint from agriculture. The implication of these figures is that those importing virtual water would have needed to use more blue water to produce the same quantity of products had they produced them locally (Mekonnen and Hoekstra 2011a). International trade in industrial products is equivalent to 4% of the global water footprint related to industrial production (Mekonnen and Hoekstra 2011b).

Fig. 13.1 Virtual water imports and exports by region. Each band represents gross virtual water export from 1995 to 1999 (Porada 2012). (With permission—see Appendix 4: Copyrights and Permissions)

13.3.1 Weighting Water to Manage Regionality

There are two main schools of thought on environmental accounting for water. The first is that every litre of water should be counted equally, whether it is sourced from a water-scarce location or not. This is the basis of the water footprint (Hoekstra and Wiedmann 2014; Mekonnen and Hoekstra 2016). The second is that water should be given a weighting factor to account for the source, i.e. 1 L of water taken from a water-scarce source might be environmentally equivalent to 2 or more litres of water taken from a water-rich source. This is the basis of the weighted water footprint (Pfister and Bayer 2014; Ridoutt and Pfister 2010, 2013).

There is no question that water bodies facing water scarcity need different management to those with an abundance of water. The idea of weighting water from different sources appeals to many.[1] However, it is an impractical solution to the problem if the goal is robust accounting of environmental currencies.

Consider Water Body A, an almost dry reservoir near Cape Town where there are severe water limitations, and Water Body B, a reservoir in Denmark with an abundance of water. It is clearly more sustainable to take a litre from Water Body B compared to Water Body A. However, the suggestion to apply a weighting factor to represent this becomes very challenging. By how much, is it preferable to extract the litre from Water Body B? Is it 10 times better? 100 times?

Ridoutt and Pfister (2013) have proposed a mechanism for water accounting with which to account for water stress of water sources using a unit of *equivalent* water (H_2Oe) to determine the weighted water footprint. Their proposal is based on a water stress index previously developed by the same authors (Pfister et al. 2009). The water stress index can range from 0.01 to 1 (it cannot be 0 in acknowledgement that every withdrawal has some impact) and is determined based on the availability of water and water withdrawals of a particular water body. One litre of H_2Oe is the burden on a water system of 1 L of water at the global average WSI.

The solution is seemingly quite elegant and appears similar to indicators such as *carbon dioxide equivalent* (CO_2e), an indicator that allows the warming impacts of different greenhouse gases to be expressed in terms of the amount of CO_2 that would produce the same amount of warming, *ozone depletion potential* (*ODP*) which indi-

---

Box 13.2 Weighted Versus Unweighted Footprints

Consider a CEO trying to reduce their weighted water footprint of her products. Her company is the only entity withdrawing water from a nearby lake, and she is aware that her company is using a large fraction of the available water. She spends substantial time and effort reducing the water consumed by her company's processes and manages to reduce total water consumption by 50%. Over the same time period, a large company moves in nearby in the same period, and this new company's withdrawals alter the water scarcity index of the water source by 50%. The CEO calculates the weighted water footprint of her products after the changes she has implemented and finds it has not changed.

The new company has a very high weighted water footprint, and this CEO is getting a lot of pressure to reduce the impacts of his products. He doesn't see an easy way to improve water efficiency within his factory, so he sabotages the first company and puts them out of business. Now that the first company is no longer taking water from the lake, his weighted water footprint drops substantially, even though his net water consumption is unchanged.

---

[1] During my extended peer community engagement, this topic was often the forefront of discussions and debate.

Box 13.3 Local Versus Global Impacts
The purpose of the Planetary Accounting Framework is to allow any scale of human activity to be compared to critical global limits. This does not preclude the need for local environmental management practices. This is not only true for water. There is high regional variation for many environmental impacts. Environmental flows which are not addressed by the Planetary Boundaries could be absolutely critical to some local ecosystems. The difference is that these are unlikely to push the balance of the Earth-system function out of a Holocene-like state. The scarcity of water in a given water body might be of critical importance locally. However, it is scarcity of water at a global scale that risks altering the function of the Earth System.

cates how much ozone will be depleted per kilogram of a substance compared to a kilogram of CFC-11, or *aerosol optical depth equivalent* (*AODe*) which is an estimation of the relative impacts of an activity on air quality. However, H_2Oe is fundamentally different. The calculation of H_2Oe is such that the weighted water footprint of an activity depends not only on the water consumed during an activity, but also by the amount of water others consume from the same water body.[2] This is inconsistent with any other environmental accounting practice. If someone releases 100 kg of methane, this is worth 250 kg of CO_2e, regardless of the activities of others. Likewise, 1 kg of nitrous oxide always has an ozone depletion potential of 0.17 ODP kg (Box 13.2).

The water scarcity index is very useful for local resource management as it gives a clear indication of the health of the water body. The problem with the weighted water footprint method is not that it is not accurate. If more users withdraw water from the same water source, the environmental impacts of each litre of water will be greater. The problem is that the method is not useful to understand and manage water impacts of consumers (see Box 13.3).

The water footprint approach where every litre of water is counted equally provides a more robust metric for water management. A company's water footprint depends solely on the company's water consumption and not on the water consumption of others. In this way, the relative water efficiency of products is apparent regardless of the source of the water. The use of this method of water accounting does not preclude the consideration of the source of water being consumed. This information could be provided in addition to the total water consumption to give a holistic view of water impacts of a given activity. This is the approach proposed here.

---

[2] The unit of H_2Oe depends on the water scarcity index which is a function of water availability and *water withdrawals*. So as total water withdrawals from a water body increase, so too do the water scarcity index and the H_2Oe of an activity.

13.4 The Indicator

The Planetary Boundary control variable for global freshwater use is *gross consumptive blue water use* (Rockström et al. 2009). The authors acknowledge that green water is a scarce resource and should be considered within the PBs. However, because of the inherent difficulty in defining a freshwater boundary that encompasses green water, they set a consumptive blue water use limit as a preliminary measure (Rockström et al. 2009; Steffen et al. 2015).

Gross consumptive blue water use is already a pressure indicator which can be scaled and applied directly to human activity. However, as shown in later in this chapter, not only water consumption but also water contamination must be considered in the Planetary Quota for water. Further, the exclusion of green water from the PQ is problematic when considering some of the potential applications of the PQs.

The Planetary Accounting Framework is not intended to be the solution for all environmental problems. It is designed to allow any scale of human activity to be compared to critical global limits. There are many local impacts that would not be considered by the Planetary Accounting Framework and thus would need to be dealt with at that scale. These impacts might be very critical to local ecosystems. The difference is that they are unlikely to push the balance of the Earth-system function out of a Holocene-like state. The scarcity of water in a local water body is one such impact. Nonetheless, there are ways in which Planetary Accounting could be used to take into account some local impacts, including water scarcity of particular water bodies. These are discussed in Chap. 18.

13.4.1 Green Versus Blue

The premise that blue water is a good indicator of total water consumption is arguable. Mekonnen and Hoekstra (2011b) show that there is a close correlation between the green water footprint of a country and total water footprint but little correlation between the total water footprint of a county and either the blue or grey water.

More importantly perhaps, when considering the exclusion of green water, is the consideration of the different purposes of the Planetary Boundaries compared to the Planetary Quotas. The Planetary Boundaries were developed to give a clear indication of overall Earth-system health. The use of a proxy indicator for total water consumption in this instance still provides an indication of water consumption compared to availability.

In contrast, the purpose of the Planetary Quotas is to be able to assess the impacts of human activity against global limits. Some argue that the use of green water, for example, to feed crops, is essentially "free" water on the basis that the water was going to fall in that area anyway. This is not an accurate account (see Box 13.4).

In planetary accounts, comparing the NZ and US timber described above, the inclusion or exclusion of green water is important. If only blue water was considered, the NZ

Box 13.4 The Need to Consider Green Water

Consider the example of a 1000-acre area in New Zealand. Originally, this area was native New Zealand forest—a very dense and damp ecosystem. The deep roots of New Zealand timbers—Kauri and Rimu—and the deep leaf litter helped the soil to retain a high moisture content. Some of this water would slowly make its way to the groundwater reserves below—well filtered by the soil and free of contaminants. The rest of the moisture would be used by the plants, and other species in the forest, and returned to the atmosphere through evapotranspiration and transpiration, ready to fall as rain again.

Today, the 1000 acres has been converted to pine forest. The rainfalls, feeding the forest. Pines have higher water uptake than Rimu or Kauri, so more water is removed from the soil, so that the soil becomes a little less moist over time. Nonetheless, the pines, like the Rimu and Kauri, transpire and release water back to the atmosphere. Then one day they reach maturity and are chopped down and removed from the site. All the rainwater that is currently held in the tree is removed from the cycle. Nutrients are applied to the soil and new trees are planted. The pines continue to dry out the soil as they absorb water more quickly than the natural level of rainfall. The degrading soil and shallower roots of the pine trees mean that water travels more quickly to the groundwater, with less filtration. Some of the natural and the added nutrients are carried away to aquifers. The soil degrades further. More pines are cut down, and more water is removed from the cycle. Over time, the soil degradation is too much, and the land becomes unsuited to forestry. The forest land is cut down and the land is converted to farmland or left as wasteland. Without trees, the water cycle changes. The rain falls, travels quickly through the soil, carrying nutrients away to local water bodies. The water bodies are starting to experience algal blooms because of all the additional nutrients carried from the soil. The soil degrades further. The grass or shrubs do not transpire as much as the trees, so less water is returned to the atmosphere. There is less moisture in the air, and therefore less rainfall.

For comparison, consider a 1000-acre area in the United States. Originally this area was grassland. The rainfall was sporadic. It is now used to grow pine. When the rain falls, there is no need to irrigate the land. But for much of the year, the land is irrigated from a water body not too far away that has a variable supply. The pine absorbs this blue water. It grows, it transpires, releasing water into the atmosphere. This falls back to the ground as rain. Some of this water makes its way back to the water body. However, over time, the water body is depleted. The pine is chopped down and taken away. All the blue water that is currently held in the tree is removed.

pine trees would have a water impact of zero, whereas the US pine trees would have a water impact >0. Yet, the amount of water used to grow these two pines is the same.

There are impacts from redirecting blue water or green water for human use. Using blue water from a water-rich source to irrigate crops may have less impact on global water scarcity than using rainfed land for crops that would otherwise have been habitat for natural ecosystems. Further, green water accounts for approximately 74% of the global average water footprint of production (Mekonnen and Hoekstra 2011a). Excluding almost 3/4 of the global water footprint from the PQ for water would give an incomplete picture.

13.4.2   Gross Water Versus Net Water

The Planetary Boundary indicator is for *gross* water consumption. This means that all water extracted from water bodies is considered, regardless of what then happens to it. Water can be borrowed from the water cycle without substantial consequence, provided it is returned in an uncontaminated state and to the same general vicinity.

Given the purpose of the Planetary Quotas, it makes more sense to consider both the extraction and the disposal of water at the end of its use. Consider, for example, two factories: Factory A and Factory B. They both produce baked beans, withdrawing the same amount of water per tin of beans from a local aquifer. Factory A dumps the waste water into the local river where it eventually makes its way, untreated, into the sea nearby. Factory B has onsite waste water treatment which treats the water to a very high standard. It is then returned to the local aquifer. The gross water consumption, the total water taken from the aquifer, is the same. However, the impacts on water use by Factory A and Factory B are not equal. The net water consumption of Factory B is the water extracted from the aquifer minus the water returned to the aquifer.

Thus, the water consumption indicator for the PQ for water is for *net* water consumption.

13.4.3   Grey Water and Novel Entities

There are hundreds of thousands of man-made chemicals, materials, and substances which have the potential to cause harm to the Earth system. The potential effects of these substances are often poorly understood. CFCs are an example of man-made chemicals that were initially thought to be a breakthrough for many human needs—in particular refrigeration—as they were so much safer than previously used refrigerants. These substances that were touted for being harmless turned out to have serious, unexpected, global effects—thinning the ozone layer to the point that every spring there is a large area with almost no ozone at all (see Chap. 11).

The use and disposal of chemicals are the two critical pressures relating the Planetary Boundary for novel entities as shown in Chap. 17. However, although novel entities are included in the Planetary Boundary framework, there is actually no limit or even a control variable proposed at this stage. The authors of the PB frame-work define novel entities as new substances, new forms of existing substances, and modified life forms that have the potential of adverse effects to the geosphere or biosphere (Steffen et al. 2015). This definition includes chemical pollution which they define as radioactive compounds, heavy metals and organic compounds devel-oped by humans, and materials or organisms engineered by humans such as nanoma-terials and plastic which can degrade to microplastics.

Of the more than 100,000 chemicals on the market (Egeghy et al. 2012), only a few thousand have toxicity data (Rockström et al. 2009). There is limited under-standing of the combined effects of these chemicals. We are still learning about the impacts of other materials such as microplastics.

There is no single indicator that covers this array of environmental impacts at a pressure level. It is difficult to imagine an indicator that could assimilate these impacts. Yet the authors have included this unitless, limitless Planetary Boundary on the basis that a global boundary for novel entities does exist. They base this premise on two rationales:

1. The direct global impact on the physiological development of humans and other organisms which changes ecosystem function or structure.
2. The indirect impacts on other Boundaries—for example, weakening species resilience to withstanding the impacts of climate change.

In the absence of a suitable indicator with which to aggregate pollutants into a com-prehensive single PB, the authors propose a twofold approach. Firstly, to focus on per-sistent pollutants that can travel long distances through the ocean or atmosphere such as mercury. The other is to identify unacceptable long-term and wide spread impacts.

The latter approach is most likely to lead to state or impact level[3] indicators. The authors deliberate over indicators such as reduced rates of or failed reproduction, neurobehavioral deficits, and compromised immune systems. The former lends itself to pressure level indicators. This approach is the basis of my approach here.

It is common practice to use water pollution as a proxy measure for chemical pollu-tion (Bjørn et al. 2014). This is typically done by determining the dilution factor—the amount of water that would be needed to assimilate any pollution. This proxy indicator does not allow for all novel entities to be considered at this stage. Entities which do not make it to water bodies, and those which cannot be diluted (e.g., plastics), are not accounted for in a water pollution metric. However, using water pollution as a proxy indicator allows for chemical pollution to be included in the Planetary Quotas in some capacity. As such, the indicator for the PQ for water includes grey water to account for chemical pollution as an interim solution. More work will be required to develop a more robust way to measure and manage novel entities.

The indicator for the PQ for water is thus *net green, blue, and grey water consumption.*

---

[3] Indicators can be classified as *states, impacts, drivers, or pressures.* A state indicator is one that describes the state of the environment. An impact indicator is one that describes a change in the state of the environment. A pressure indicator is one that describes flows to the environment. Pressure indicators are the type of indicators used for the Planetary Quotas. See Chap. 7 for more details.

13.5 The Limit

The Planetary Boundary for freshwater use is <4000 km³/year of gross consumptive blue water use with an uncertainty zone of 4000–6000 km³/year (Rockström et al. 2009). Consumptive use of blue water is about 2600 km³/year (Steffen et al. 2015). It has been estimated that approximately 25–50% more blue water may be needed 2050 to ensure food security (Molden 2007).

There is little agreement in the literature as to a global limit for net green, blue, and grey water consumption. It has been estimated that as much as 90% of green water flows (Rockström et al. 1999) and 20–50% of blue-water flows (Smakhtin 2008) are required to maintain ecosystems (including rainfed croplands). Global green water availability is about 70,000 km³/year, and blue water, about 12,500 km³/year (Postel et al. 1996), so these limits would indicate that approximately 7000 km³/year of green water and a further 5000 km³/year of blue water could be consumed by humans—a total of 12,000 km³/year. Other authors suggest global blue water scarcity will be reached when withdrawals exceed 5000–6000 km³/year (Raskin et al. 1997; Vörösmarty et al. 2000; De Fraiture et al. 2001).

On the basis that more than 30% of major groundwater sources are currently being depleted, Hoekstra (2017) argues that we are already at the Boundary, if not beyond it, and that a precautionary approach would be to set the limit no higher than current net global water consumption ≈8500 km³/year (Hoekstra 2017). Annual gross blue water consumption is approximately 2600 km³/year. In contrast, annual net blue water consumption is approximately 1000 km³/year (derived from (Mekonnen and Hoekstra 2011b)). This would leave a quota of approximately 6400 km³/year of green water, slightly below the maximum appropriation of green water proposed by (Rockström et al. 1999). The remaining 1100 km³/year would be available as grey water to assimilate pollutants.

In the absence of an alternative basis for the limit for water consumption, the Planetary Quota for water is, thus, net green, blue, and grey water consumption ≤8500 km³/year.

This PQ can be compared to the water footprint as defined by Mekonnen and Hoekstra (2011b) for any scale of activity.

13.6 Discussion

The water footprint of the global average consumer between 1996 and 2005 was 1385 m³/year. 92% of this was from agricultural products, 5% industrial goods, and 4% for domestic water use (Mekonnen and Hoekstra 2011b). If everyone consumed the global average amount of water, the global water footprint at today's population would be over 10,000 km³. Yet, approximately 780 million people do not have access to clean water and 2.5 million do not have access to sanitation (WWF 2014).

Box 13.5 Eating Water

It is interesting to note that of the total global water footprint, only 4% was for domestic water use. 92% was for agricultural products; nearly a third of this was related to the production of animal products. The consumption of meat accounts for 22% of the water footprint of the average consumer. The average WF per calorie of beef is 20 times larger than cereals and starchy roots (Mekonnen and Hoekstra 2011a). Diet is thus one of the greatest contributors to the global water footprint.

Table 13.1 Examples of different scales of activity which have or could contribute to achieving the PQ for water consumption

Achieving the Planetary Quota for water consumption			
	Community	**Government**	**Business**
Large	Develop global organizations dedicated to the management of global water resources, e.g. *the Global Water Initiative*	Develop a global treaty for water management	Develop innovative low water technologies
Medium		Develop holistic national water strategies, e.g. *Singapore is targeting the collection of every drop of water, the endless reuse of water, and desalination of sea water to meet national water demands*	Relocate water-intensive activities to locations where water bodies are not suffering from water scarcity
Small	Install household or community grey water recycling systems	Set local irrigation limit, e.g. *local councils in Perth, Western Australia, set irrigation rules such as the days that watering is allowed, and the number of minutes plants can be watered*	Choose water-efficient raw materials for products
Individual	Eat a plant-based diet		Educate staff on water footprints

Falkenmark (1986) estimates that approximately 500 m^3/p/year is needed to run a modern society. At the current population, this would give a global water footprint of approximately 3800 km^3/year. Even at a population of nine billion, this gives a total footprint of 4500 km^3/year. Both estimates are within the Planetary Quota for water (Box 13.5).

Table 13.1 lists examples of activities at different scales to show how a poly-scalar approach to managing global water consumption might work at different scales of activity.

13.7 Conclusion

Water is a life essential and irreplaceable resource. Over 1/3 of major groundwater aquifers are currently being depleted, suggesting that we are already consuming more than the planet's capacity.

The Planetary Boundary for water is in the unit gross, blue water consumption. The proposed PQ indicator is in net green, blue, and grey water consumption. The different indicator is to allow for more robust comparison of different human activities, to accommodate water management through water treatment strategies, and to incorporate the Planetary Boundary for novel entities.

The Planetary Quota for water is net blue, green, and grey water consumption ≤ 8500 km³/year. This can be compared to the water footprint of any scale of activity. The limit is set on the basis of the current global water footprint.

References

AMTA (2016) Membrane desalination power usage put in perspective. America's Authority in Membrane Treatment, Stuart, FL

Bjørn A, Diamond M, Birkved M, Hauschild MZ (2014) Chemical footprint method for improved communication of freshwater ecotoxicity impacts in the context of ecological limits. Environ Sci Technol 48:13253

De Fraiture C, Molden D, Amarasinghe U, Makin I (2001) PODIUM: projecting water supply and demand for food production in 2025. Phys Chem Earth B Hydrol Oceans Atmos 26:869–876

Egeghy PP, Judson R, Gangwal S, Mosher S, Smith D, Vail J, Cohen Hubal EA (2012) The exposure data landscape for manufactured chemicals. Sci Total Environ 414:159–166

Falkenmark M (1986) Fresh water—time for a modified approach. Ambio 15:192–200

Hoekstra AY (2017) Water footprint assessment: evolvement of a new research field. Water Res Manag 31:1–21

Hoekstra AY, Wiedmann TO (2014) Humanity's unsustainable environmental footprint. Science 344:1114

Leahy S (2018) From not enough to too much, the world's water crisis explained. National Geographic, National Geographic Society

Mekonnen MM, Hoekstra AY (2011a) The green, blue and grey water footprint of crops and derived crop products. Hydrol Earth Syst Sci 15:1577–1600

Mekonnen MM, Hoekstra AY (2011b) National water footprint accounts: the green, blue and grey water footprint of production and consumption. Value of water research report series no. 50. UNESCO-IHE, Delft

Mekonnen MM, Hoekstra AY (2016) Four billion people facing severe water scarcity. Sci Adv 2:e1500323

Molden D (2007) Comprehensive assessment of water management in agriculture. International Water Management Institute, Earthscan, London

Oki T, Kanae S (2006) Global hydrological cycles and world water resources. Science 313:1068–1072

Pfister S, Bayer P (2014) Monthly water stress: spatially and temporally explicit consumptive water footprint of global crop production. J Clean Prod 73:52–62

Pfister S, Koehler A, Hellweg S (2009) Assessing the environmental impacts of freshwater consumption in LCA. Environ Sci Technol 43:4098

Pitt C (2018) Cape Town water consumption drops by 43 million litres. *news24*. news24

Porada B (2012) Virtual water imports and exports by region

Postel SL, Daily GC, Ehrlich PR (1996) Human appropriation of renewable fresh water. Science 271:785–788

Raskin P, Gleick P, Kirshen P, Pontius GA, Strzepek K (1997) Water futures: assessment of long-range patterns and problems. In: SEI (ed) Comprehensive assessment of the freshwater resources of the world. SEI, Stockholm

Ridoutt BG, Huang J (2012) Environmental relevance—the key to understanding water footprints. Proc Natl Acad Sci U S A 109:E1424; author reply E1425

Ridoutt BG, Pfister S (2010) A revised approach to water footprinting to make transparent the impacts of consumption and production on global freshwater scarcity. Glob Environ Chang 20:113–120

Ridoutt B, Pfister S (2013) A new water footprint calculation method integrating consumptive and degradative water use into a single stand-alone weighted indicator. Int J Life Cycle Assess 18:204–207

Rockström J, Gordon L, Folke C, Falkenmark M, Engwall M (1999) Linkages among water vapor flows, food production, and terrestrial ecosystem services. Ecol Soc 3:5

Rockström J, Steffen W, Noone K, Persson A, Chapin FS, Lambin E, Lenton TM, Scheffer M, Folke C, Schellnhuber HJ, Nykvist B, De Wit CA, Hughes T, Van der Leeuw S, Rodhe H, Sorlin S, Snyder PK, Costanza R, Svedin U, Falkenmark M, Karlberg L, Corell RW, Fabry VJ, Hansen J, Walker B, Liverman D, Richardson K, Crutzen P, Foley J (2009) Planetary boundaries: exploring the safe operating space for humanity. Ecol Soc 14:32

Secretariat of the CBD (2001) Global biodiversity outlook 1. In: SotCoB (ed) Diversity. UNEP, CBD, Montreal

Shiklomanov IA (1993) Water in crisis: a guide to the world's fresh water resources. In: Gleick P (ed) Oxford University Press, New York

Smakhtin VU (2008) Basin closure and environmental flow requirements. Int J Water Res Develop 24:227–233

Steffen W, Richardson K, Rockström J, Cornell SE, Fetzer I, Bennett EM, Biggs R, Carpenter SR, De Vries W, De Wit CA, Folke C, Gerten D, Heinke J, Mace GM, Persson LM, Ramanathan V, Reyers B, Sörlin S (2015) Planetary boundaries: guiding human development on a changing planet. Science 347:1259855

UN (2015) The human right to water and sanitation. UN, Geneva

Vörösmarty CJ, Green P, Salisbury J, Lammers RB (2000) Global water resources: vulnerability from climate change and population growth. Science 289:284–288

Wangnick Consulting (1990) IDA worldwide desalting plants inventory. International Desalination Association, Englewood, NJ

WWF, WWL (2014) Freshwater: what's at stake, what we're missing, what we're losing, what it's worth. http://wwf.panda.org/about_our_earth/about_freshwater/importance_value/. Accessed 27 Mar 2014

Chapter 14
A Quota for Nitrogen

All we are is a lot of talking nitrogen
Arthur Miller

Abstract Reactive nitrogen is necessary to grow food. It is often the limiting factor for plant growth, and without it, farming yields would be substantially lower. However, the overuse of nitrogen fertilizers has led to high levels of nutrient run off, causing algal blooms and therefore anaerobic dead zones in rivers, lakes, and oceans.

The Planetary Boundary for nitrogen is a maximum of 62 Tg N/year of intentionally fixated nitrogen. This indicator is scalable, but not easily comparable to human activity. Further, it does not consider downstream denitrification processes that can reduce the environmental impacts of nitrogen use.

The Planetary Quota indicator for nitrogen is *net nitrogen consumed* ≤ 62 Tg N. This includes virtual nitrogen that is lost to the environment during the production of food and accounts for the removal and recycling of nitrogen from the human nitrogen cycle. The limit is based on the premise that the Planetary Boundary value is based on the maximum flow of nitrogen to waterways. Net nitrogen consumed will eventually end in waterways. This limit can be compared to the nitrogen footprint of any scale of human activity. Current annual nitrogen consumption exceeds the PQ for nitrogen.

14.1 Introduction

Nitrogen is the most prevalent element in the Earth's atmosphere. Approximately 78% of the atmosphere (by volume) is nitrogen gas (N_2). Nitrogen is one of the fundamental building blocks of life. It is in chlorophyll, the green pigment in plants that is responsible for photosynthesis, is a building block of protein, and is critical to other cellular elements that are essential to life (Wagner 2011). However, in its most abundant form, a stable gas, it cannot be used by most living organisms. Reactive nitrogen (N_r) is the form of nitrogen that is needed for life. In contrast to

© Springer Nature Singapore Pte Ltd. 2020
K. Meyer, P. Newman, *Planetary Accounting*,
https://doi.org/10.1007/978-981-15-1443-2_14

nitrogen gas, reactive nitrogen is relatively scarce. A lack of available reactive nitrogen is often the limiting factor for natural ecosystems. This can also be the limiting
factor for intentional human ecosystems (e.g. farms).

Excessive loss of reactive nitrogen to the environment can have harmful
impacts including eutrophication, smog, acid rain (which harms plant and aquatic
life and infrastructure), and stratospheric ozone depletion (n-print 2011). Nitrous
dioxide (one form of reactive nitrogen) is also an important greenhouse gas (GHG)
(see Chap. 9).

Reactive nitrogen is considered within several PQs, not only the PQ for nitrogen.
This is because of the different impacts of reactive nitrogen on the Earth system.
The PQ for nitrogen addresses the impacts of nitrogen on water bodies, i.e. eutrophication. The impacts of reactive nitrogen as a greenhouse gas, and ozone depleting substance, and an aerosol precursor are addressed in the PQs for methane and
nitrous oxide and aerosols, respectively (see Chaps. 9 and 11).

The inclusion of reactive nitrogen across several PQs does not constitute double
counting for the purpose of planetary accounting. This is because each PQ must be
respected. There is no mechanism with which to amalgamate the PQs into a single
indicator or to offset one against another. However, if the PQs were going to be used
as the basis for a tax scheme, it would be important to include a mechanism so that
excess reactive nitrogen use was not charged more than once (see Chap. 18).

This chapter begins with an introduction to the nitrogen cycle, human use of
nitrogen, and the critical environmental impacts from reactive nitrogen. The PQ
indicator *net reactive nitrogen released to the environment* is presented, and the case
for the preliminary limit for nitrogen is made. The chapter concludes with a discussion of the PQ for nitrogen and the types of actions that might be needed in order for
humanity to live within this PQ.

14.2   Background

14.2.1   The Natural Nitrogen Cycle

As discussed in Chap. 8, the Earth system has natural biogeochemical cycles in
which a chemical substance moves between the atmosphere, biosphere (life on
Earth), lithosphere (Earth's crust), and hydrosphere (surface and atmospheric
water). Nitrogen is one of the substances which moves through such a cycle. There
are five key processes in the nitrogen cycle: fixation, ammonification, nitrification,
assimilation, and denitrification.

14.2.1.1 Nitrogen Fixation

The most common form of nitrogen is N_2, two nitrogen atoms bonded together. This is very stable and unusable by plants. Nitrogen fixation is the process of converting nitrogen from this stable nitrogen gas form to useful forms ammonia (NH_3) and ammonium (NH_4). Natural nitrogen fixation is generally done by nitrogen-fixing bacteria through a metabolic process that is similar to the way humans and other animals convert oxygen (O_2) to carbon dioxide (CO_2) when we breathe. The bacteria can be free-living in the soil or water, can be associated with plants (typically grasses—including rice, wheat, corn, oats, and barely), or can have a symbiotic relationship with plants (typically legumes such as alfalfa, beans, clover, peanuts, and soybeans). This biological fixation accounts for 90% of natural reactive nitrogen in terrestrial ecosystems.

The other natural form of nitrogen fixation occurs when high levels of energy are applied to nitrogen gas which breaks apart the nitrogen molecules, leaving them ready to make new bonds. The high energy can come from lightening, forest fires, and the heat from volcanic eruptions. Oxidized forms of nitrogen are produced in the atmosphere (NO_x), and then this settles to Earth's surface where it can be used assimilated by plants.

Together terrestrial ecosystems are estimated to release approximately 65 Tg N/year and marine biological systems a further 140 Tg N/year.

14.2.1.2 Nitrification

Bacteria in the soil convert ammonium (NH_4) and ammonia (NH_3) to nitrite (NO_2) and then to nitrate (NO_3). This process typically occurs aerobically. It is done exclusively by prokaryotes which are single-celled bacteria and cyanobacteria without a nucleus or membrane.

14.2.1.3 Assimilation

Assimilation is the absorption of ammonia or nitrate from soil by plants. Plants convert nitrate to nitrite ions and ammonium ions—the forms needed to become amino acids, nucleic acids, proteins, and chlorophyll. Animals (and humans) get their nitrogen from these plant tissues.

14.2.1.4 Ammonification

Ammonification is the reverse of assimilation. The organic nitrogen (proteins, acids) is converted back into ammonia. When plants and animals defecate, urinate, or die, the organic nitrogen is available to bacteria and fungi which can return it to ammonia. This ammonia is left in the environment ready for return through the cycle via nitrification or assimilation.

14.2.1.5   Denitrification

Nitrates and nitrites are converted back into nitrogen gas by bacteria in anaerobic conditions such as deep in the soil or near the water table. Wetlands are a very important part of the denitrification process. Denitrifying bacteria release nitrous oxide as well as nitrogen gas back into the atmosphere.

14.2.2   Human Use of Nitrogen

The management of nutrients in soil (including nitrogen) can be traced back in history to as early as 6000 BC—when Middle Eastern farmers practiced crop rotation. The Bible has reference to a "Sabbath of the Land" which meant that every 7 years they would leave the land return to its natural state. Farming practices developed over the years from two-field rotation where only half of the land was farmed each year and the other half left to recover to a three-field system, where two crops would be rotated both seasonally and annually and a third of the land would be rested every year. Four-field rotation began in the early sixteenth century—this included seasonal rotation, annual rotation, and, importantly, rotated arable and livestock farming. The rotations all included leguminous and cereal crops which produced ammonia in the soil for the other crops. The addition of a livestock rotation increased the return of nitrogen to the soil through animal urine and faeces.

Biological N-fixation is slow and limited, so in addition to crop rotations, the use of natural fertilizers was common. Manure, guano (bird droppings), and human waste, all of which are rich in nitrogen, were applied to fields to promote plant growth.

Throughout this time, farmers were reaping the benefits of well-managed nutrients in the soil but without understanding the chemistry behind their actions. In 1840 Justus von Liebig discovered the important roles ammonia (one of the reactive forms of nitrogen, and later of phosphorus) (discussed in Chap. 15) (Liebig 1840). After his discovery, nitre mining for potassium nitrate for use as a fertilizer became common. However, at the beginning of the twentieth century, there were concerns that the demand for nitre would quickly outstrip the supply and research into sources of ammonia increased.

14.2.2.1   The Haber-Bosch Process and the Green Revolution

In 1909 Fritz Haber discovered a way to convert nitrogen gas into ammonia. He placed hydrogen and nitrogen gas under high pressure to force a chemical reaction that converted them to ammonia (NH_3). The Baden Aniline and Soda Factory

(BASF), a German chemical company, bought the process from Haber and assigned employee Carl Bosch to the job of scaling Haber's process up to an industrial scale. Bosch succeeded in 1910 and the procedure became known as the Haber-Bosch process.

The Haber-Bosch process occurred at a similar time to the start of phosphorous mining. This early twentieth-century period is thus known as the Green Revolution. Agricultural production grew exponentially, as did population growth. It is estimated that without the Haber-Bosch process, only three billion people could be fed given current diets and agricultural practices (Erisman et al. 2008). Between 1900 and 2000, the population quadrupled, yet the agricultural area used to feed the global population only increased by 30% (De Vries et al. 2013). Nitrogen and phosphorus fertilizer was not the only reason for the improvement in agricultural yield which allowed this to happen. Plant breeding, herbicides, and pesticides were also important factors. However, the newfound ability of humans to intentionally produce reactive nitrogen is one of the most important factors (De Vries et al. 2013). In this period there was a 50-fold increase in nitrogen fertilizers. De Vries et al. (2013) postulate that without fertilizer, a similar population growth would have required a proportional increase in agricultural area and thus other major environmental impacts such as high levels of biodiversity loss.

The Haber-Bosch process is still the primary method for developing nitrogen fertilizer used today. Ammonia is in fact one of the most highly produced inorganic chemicals. Projections are that more than 187 million tonnes of nitrogen fertilizer will be used in 2018 (FAO 2017).

14.2.3   The Impacts of Nitrogen

Without human interference, approximately 0.5 kg of N/ha/year is deposited (Galloway et al. 2008). Now, for many places, average deposition is >10 kg N/ha/year (Science Communication Unit 2013).

Approximately 75% of man-made reactive nitrogen is from N-fixation and the remaining 25% from fossil-fuel and biomass burning. All reactive nitrogen created through fossil-fuel combustion is lost to the environment (Leach et al. 2012). Most reactive nitrogen used in agriculture is lost to the air, soil, or water. Only a small proportion of nitrogen applied to agriculture is taken up by crops. Humans and other species do not absorb nitrogen, so all of the nitrogen taken up by crops and then consumed by livestock and people is expelled in urine and faeces. In the case of livestock, some of this is returned to the natural nitrogen cycle. Before monocultural agriculture, this release of nutrients from livestock was how much of the land was fertilized. However, intensive monocultural grazing means that excessive levels of nitrogen are released to the environment—more than can be absorbed by the natural cycle.

Human waste used to be returned to land and the nutrients returned to their natural cycle. It is now released into water. It is possible to denitrify waste water—a process that removes approximately 90% of nitrogen from sewage. However, only a small proportion of global sewage is treated. Most of this nitrogen is released back into the environment.

Denitrifying bacteria not only produce nitrogen gas; they also produce nitrous oxide (N_2O), a dangerous greenhouse gas. The use of nitrogen fertilizer has led to large increases in the amount of N_2O released into the atmosphere from agriculture.

Total reactive nitrogen production in agriculture is more than double the pre-industrial natural amount in terrestrial ecosystems (Science Communication Unit 2013). The nitrogen used in modern agriculture is leading to widespread environmental change (Rockström et al. 2009). Human activity is altering the natural nitrogen cycle.

There are many local but also global consequences from the use of human fixated nitrogen. Excessive use of nitrogen in agriculture leads to eutrophication of terrestrial ecosystems (De Vries et al. 2013). Eutrophication is excessive nutrient richness which can cause high growth of plants such as algae—known as algal blooms—which in turn prevent oxygen and sunlight from reaching the water below. This can lead to hypoxic conditions, wiping out fish and other aquatic species. The die-off of algal blooms releases toxins into the water which can further reduce biodiversity in the area (De Vries et al. 2013). This can change the function of the ecosystems and reduce biodiversity (De Vries et al. 2013).

Nitrogen can cause acidification of soil and water. Airborne reactive nitrogen is one of the primary causes of acid rain. There are excessive nitrates in much of the world's drinking water which has negative health impacts (De Vries et al. 2013). Airborne nitrogen particles are dangerous to human health and crop yields (De Vries et al. 2013). Nitrogen dioxide (NO_2) is the dominant source of oxygen atoms for toxic, ground level ozone (O_3), while nitrous oxide leads to the depletion of the important layer of stratospheric ozone. Nitrous dioxide is one of the critical greenhouse gases that is causing climate change. Nitrous oxide also leads to stratospheric ozone depletion (Science Communication Unit 2013).

14.3 The Indicator

The existence of a global limit for nitrogen is debated in the literature on the basis that nitrogen impacts are location specific. In fact, some of the impacts are globally dispersed, for example, the emissions of nitrous oxide from the use of nitrogen fertilizer. Moreover, the location of nitrogen use is often spatially distant to the location of the end use. The concept of virtual nitrogen, i.e. the nitrogen used in the production of products [similar to the concept virtual water (see Chap. 14)], allows us to better see the global distribution of a regional or local problem.

The Planetary Boundary indicator for nitrogen, the industrial and intentional biological fixation of nitrogen, is a pressure.[1] However, it is not a pressure that suits the requirements of the Planetary Quotas. It is very difficult to link the fixation of nitrogen to downstream activities at different scales. A more scalable and applicable indicator would be the amount of fixated nitrogen used and lost to the environment.

A nitrogen footprint (NF) has been developed to measure reactive nitrogen used in human activities (Leach et al. 2012). This indicator assesses the net nitrogen released to the environment by human activity. The nitrogen considered is both direct nitrogen consumed, i.e. the nitrogen in the carrot or steak that a person is eating, and the virtual nitrogen. Virtual nitrogen is the nitrogen that has been lost to the environment downstream. It includes the ammonia lost to the groundwater when growing the carrot, the nitrogen released in urine and manure before the cow was taken to the slaughter house, and the nitrous oxide emissions released from the burning of fossil fuels to transport the carrot and steak to the supermarket and then the person's house.

It is possible to remove nitrogen from waste water—industrial denitrification. As much as 90% of the nitrogen in sewage can be removed before the waste is released to the environment. The nitrogen footprint includes a mechanism to account for this positive behaviour by removing this amount from the footprint in the instance where wastewater is treated.

The control variable used to assess nitrogen footprints is the *net reactive nitrogen released to the environment*. Unlike the Planetary-Boundary indicator, this Pressure indicator can be related directly to any human activity, as shown by its use in determining nitrogen footprints of people, products, and nations (Leach et al. 2012; Pierer et al. 2014).

14.4 The Limit

The PB limit for nitrogen is set at a point estimated to limit the impacts of agricultural nitrogen on the environment while still meeting the world's need for food (De Vries et al. 2013). De Vries et al. (2013) assessed critical environmental limits for ammonia in the air, nitrous dioxide in the air, and nitrogen in surface runoff. They then estimated the minimum amount of nitrogen fertilizer needed to feed a future population of nine billion people. They conclude from these assessments that an appropriate boundary would be a fixation rate of 62–100 Tg N/year. The authors of the Planetary Boundaries updated the limit to 62 Tg N/year, the most stringent end of the range.

This limit is lower than the estimated minimum nitrogen that would need to be fixated to feed the population at current average nitrogen use efficiency (the

---

[1]Environmental indicators can be classed as States, Impacts, Drivers, or Pressures under the European Union DPSIR framework. Pressures describe flows to the environment and are the type of indicator used for the Planetary Quotas. See Chap. 5 for details.

amount of nitrogen taken up by different plants) of 80 Tg N/year. However, the authors estimated that minimum N-fixation could drop to 50 Tg N/year with a nitrogen efficiency increase of 25%, an efficiency increase they deemed to be feasible (De Vries et al. 2013).

The globally *intended nitrogen fixation* (the PB indicator) is not equivalent to the *net reactive nitrogen released to the environment* (the PQ indicator). However, the basis of the PB limit of 62 Tg/N (i.e. the maximum amount of N_r that can safely be released to the environment) is also an appropriate basis for the PQ indicator limit. As such, the PQ for nitrogen is net reactive nitrogen released to the environment $\leq$62 Tg N/year.

14.5   Discussion

There are currently approximately 112 Tg N/year released to the environment (derived from Steffen et al. 2015; Keeler et al. 2016), almost double the PQ for nitrogen. The authors of the PB framework suggest that the PB for nitrogen could be met with improved farming practices and innovations such as the use of human waste onto productive landscapes (Rockström et al. 2009).

A study of potential reactive nitrogen reductions in the UK showed that reductions of up to 63% were possible leading to a per capita N-footprint of 10 kg N/ person/year (Stevens et al. 2014). Based on the current global population, and equal per capita nitrogen, the Quota equates to approximately 8.3 t N/year. At a future population of nine billion, this would reduce to 6.9 t N/person/year.

Different foods have different nitrogen uptake efficiency (NUE). The higher the uptake, the less nitrogen lost before the food is consumed. The average NUE for animal proteins is very low—at about 8% (meaning that 92% of the nitrogen used to develop the food is lost to the environment before the food is consumed). Plant-based food has an average NUE of 20%. The reason for the low efficiency in animal proteins is that nitrogen is lost both in the growing of the animal fodder and in the animal waste (manure).

Table 14.1 shows examples of the sorts of activities that might be different across different scales and sectors in order to live within the PQ for nitrogen.

14.6   Conclusions

Nitrogen is a critical element in food production. However, too much nitrogen can cause run-off of nitrogen into water ways. This can lead to algal blooms which can be very harmful to aquatic eco-systems.

The Planetary Quota for nitrogen is net nitrogen released to the environment through agriculture. This is to differentiate it from nitrogen emissions through

Table 14.1 Examples of different scales of activity which have or could contribute to achieving the PQ for nitrogen

Achieving the Planetary Quota for nitrogen

	Community	Government	Business
Large	Develop a global organization dedicated to the measurement of nitrogen use, *e.g. n-print*	Develop a global agreement for nitrogen management	Develop innovative solutions to limit the release of nitrogen to the environment
Medium		Set maximum national nitrogen application rates	
Small	Community compost initiatives, *e.g. Compost Revolution— Australia's largest community of composters and worm farmers comprising more than 30,000 households in Sydney*		Alter farming practices to include on-farm nitrogen cycling with manure from livestock to feed crops
Individual	Eat an organic and plant-based diet		Start a business to manage challenges of composting: *e.g. Steve Rickerby realized that office buildings were not composting due to lack of space to compost on-site so started a business collecting compostable from offices, food courts, schools, universities, hotels, and cafes*

burning fossil fuels, for example, which are captured in the PQ for methane and nitrous oxide and the PQ for aerosols.

The limit for the PQ for nitrogen is the release of reactive nitrogen to the environment ≤ 62 Tg N. This can be compared to the nitrogen footprint of any scale of human activity. The limit is based on the Planetary Boundary limit for maximum global nitrogen fixation.

References

De Vries W, Kros J, Kroeze C, Seitzinger SP (2013) Assessing planetary and regional nitrogen boundaries related to food security and adverse environmental impacts. Curr Opin Environ Sustain 5:392–402

Erisman JW, Sutton MA, Galloway J, Klimont Z, Winiwarter W (2008) How a century of ammonia synthesis changed the world. Nat Geosci 1:636

FAO (2017) World fertilizer trends and outlook to 2020. Food and Agriculture Organization (FAO) of the United Nations, Rome

Galloway JN, Townsend AR, Erisman JW, Bekunda M, Cai Z, Freney JR, Martinelli LA, Seitzinger
    SP, Sutton MA (2008) Transformation of the nitrogen cycle: recent trends, questions, and
    potential solutions. Science 320:889–892
Keeler BL, Gourevitch JD, Polasky S, Isbell F, Tessum CW, Hill JD, Marshall JD (2016) The
    social costs of nitrogen. Sci Adv 2:e1600219
Leach AM, Galloway JN, Bleeker A, Erisman JW, Kohn R, Kitzes J (2012) A nitrogen footprint
    model to help consumers understand their role in nitrogen losses to the environment. Environ
    Dev 1:40–66
Liebig JFV (1840) Die organische Chemie in ihrer Anwendung auf Agricultur und Physiologie
    (Organic chemistry in its applications to agriculture and physiology). F. Vieweg, Braunschweig
N-Print (2011) Background information [Online]. n-print. http://www.n-print.org/node/31
Pierer M, Winiwarter W, Leach AM, Galloway JN (2014) The nitrogen footprint of food products
    and general consumption patterns in Austria. Food Policy 49:128–136
Rockström J, Steffen W, Noone K, Persson A, Chapin FS, Lambin E, Lenton TM, Scheffer M,
    Folke C, Schellnhuber HJ, Nykvist B, De Wit CA, Hughes T, Van Der Leeuw S, Rodhe H,
    Sorlin S, Snyder PK, Costanza R, Svedin U, Falkenmark M, Karlberg L, Corell RW, Fabry VJ,
    Hansen J, Walker B, Liverman D, Richardson K, Crutzen P, Foley J (2009) Planetary boundar-
    ies: exploring the safe operating space for humanity. Ecol Soc 14:32
Science Communication Unit (2013) Science for environment policy in-depth report nitro-
    gen pollution and the European environment—implications for air quality policy. European
    Commission, Bristol
Steffen W, Richardson K, Rockström J, Cornell SE, Fetzer I, Bennett EM, Biggs R, Carpenter SR,
    De Vries W, De Wit CA, Folke C, Gerten D, Heinke J, Mace GM, Persson LM, Ramanathan
    V, Reyers B, Sörlin S (2015) Planetary boundaries: guiding human development on a changing
    planet. Science 347:1259855
Stevens CJ, Leach AM, Dale S, Galloway JN (2014) Personal nitrogen footprint tool for the United
    Kingdom. Environ Sci Process Impacts 16:1563–1569
Wagner SC (2011) Biological nitrogen fixation. Nat Edu Know 3:15

Chapter 15
The Phosphorus Quota

> *Life can multiply until all the phosphorus has gone and then*
> *there is an inexorable halt which nothing can prevent*
> *Isaac Asimov (1974)*

Abstract Phosphorus is a chemical element that is vital to all life on Earth. It is critical in the formation of genetic instructions, in the production of cells, in providing internal energy to power cellular function, and in the formation of seeds and fruit.

Before human interference, the phosphorous cycle was in balance. Phosphorus consumed by plants and animals was returned to the soil. Waterways transported phosphorus as needed for aquatic life. A slow weathering of phosphate rocks was matched by the slow formation of new rocks in phosphorus-rich ocean sediments.

Since the Industrial Revolution, humans have altered the phosphorous cycle. Humans are extracting millions of tonnes of mineral phosphate from rocks every year. This is applied to land as fertilizer to grow food, and then much of it is released as waste to waterways. There is some concern as to the level of remaining reserves of phosphate rock and whether we are likely to run out of this critical resource in the near term. However, the reason for the inclusion of phosphorus in the Planetary Boundaries is not the potential supply shortfall but rather the potential environmental impacts. The excessive release of phosphorus to water can lead to algal blooms and thus anoxic events, wiping out the entire ecosystems. This process is believed to have happened on a global scale in the past—creating anoxic oceans and driving a global mass extinction of marine life.

The Planetary Quota for phosphorus is 11 Gt/year released to the environment. This is based on the Planetary Boundary for maximum flow of phosphorus to the sea. The limit can be compared to phosphorus released during any scale of human activity.

© Springer Nature Singapore Pte Ltd. 2020
K. Meyer, P. Newman, *Planetary Accounting*,
https://doi.org/10.1007/978-981-15-1443-2_15

15.1   Introduction

The human-induced alteration of the phosphorous cycle is perhaps one of the least well-known global environmental crises facing us today. Phosphorus is a chemical element that is essential to all life on Earth. It is the 11th most abundant element in the Earth's crust (Schröder et al. 2010). Phosphorus is an essential component of genetic material—DNA (deoxyribonucleic acid) and RNA (ribonucleic acid). It is necessary for the production of cell membranes and for the creation of seeds and fruit which are fundamental to the life cycle of fauna. All living organisms need phosphorus every day to produce energy. Even bacteria need phosphorus to survive (Ashley et al. 2011). There is no alternative to phosphorus. There is no synthetic substitute (Science Communication Unit 2013).

Phosphorus is very reactive so it is unusual to find it in its pure form. The most common form of available phosphorus is when it occurs as a phosphate (PO^{3-}_4) in rocks, natural sediments, and manure called guano. Phosphates are the backbone of DNA. They are also a key component of adenosine triphosphate (ATP), an important chemical which transfers energy (Ashley et al. 2011).

The name phosphorus comes from the Greek words *phôs* which means *light* and *phoros* which means *bearer*. Pure phosphorus glows in the dark (Ashley et al. 2011) and can sometimes be seen on the ocean surface at night. Humans mine phosphorus from phosphate rock and deposits for a variety of applications. The predominant use, which accounts for 90% of mined phosphorus, is for fertilizer and animal feed (Prud'homme 2010). A further 7% is used in detergents, although this is declining as most high-income countries do not allow its use in detergents anymore (Liu et al. 2008). The small remaining amount of phosphorus is used as flame retardant, for metal surface treatment, and in ceramic production (Liu et al. 2008).

This chapter begins with three sections that provide a background to phosphorus and its use: an introduction to phosphorus and the phosphorous cycle, an overview of the history of human appropriation of phosphorus, and a discussion of phosphorus as a nonrenewable resource. The second part of the chapter is about the Planetary Quota (PQ) for phosphorus. The chapter concludes with a discussion on how our current phosphorous use compares with the PQ for phosphorus and what living within this PQ might mean practically.

15.2   The Phosphorus Cycle

Phosphorus is one of the chemical substances that moves between Earth's biosphere (life on Earth), lithosphere (Earth's crust), atmosphere (the layer of gases surrounding the Earth), and the hydrosphere (surface and atmospheric water) naturally in one of Earth's biogeochemical cycles (see Chap. 8).

The phosphorous cycle can be broken down into three sub-cycles:

1. The inorganic phosphorous cycle: Phosphorus is accumulated as sediment on the seafloor. Over millions of years, the sediment is turned into rock through geological pressure. Tectonic shifts under Earth's crust expose phosphate rock. The exposed rocks are subject to weathering, releasing phosphorus back into the environment (Föllmi 1996; Schlesinger and Bernhardt 2013).
2. The land-based organic phosphorous cycle: Plants take phosphorus from the soil. Plants are either eaten by animals, in which case the phosphorus is returned to the soil via urine and faeces, or when the plants die and decay so that the phosphorus returns to the soil directly. It takes an average of 1 year for a molecule of phosphorus to complete this cycle (Liu et al. 2008).
3. The water-based organic phosphorous cycle: Phosphorus is circulated between creatures in lakes, rivers, and oceans. This is the most rapid cycle—it takes only weeks for a molecule to complete the cycle (Liu et al. 2008).

Unlike carbon, nitrogen, water, and oxygen cycles, the phosphorous cycle does not include a gaseous phase. As such, there is no atmospheric link between the land and the ocean other than the wind transport of phosphorus containing soil or water particles (Liu et al. 2008).

Before humans began mining phosphate rock, human use of phosphorus did not disturb the balance of the natural phosphorous cycle. However, over a relatively short period, humans have extracted hundreds of millions of tonnes of mineral phosphorus from the inorganic phosphorus and released it into the organic phosphorous cycles with severe environmental consequences.

15.3 Human Use of Phosphorus: A Brief History

Phosphorus has a speckled history full of amazement and danger from the accidental discovery of phosphorus when burning urine to the use of phosphorus as a weapon. The use of phosphorus to improve crop yield has occurred unwittingly for over 40,000 years. The aboriginal "firestick" farming—burning of sections of forest to promote agricultural growth—was effective because the phosphorus in the ash was temporarily available for plants in otherwise very phosphorous poor soil (Cordell 2001; Flannery 1994) (see Fig. 15.1).

China used human waste (known as "night soil") to fertilize land as early as 5000 years ago (Ashley et al. 2011). Medieval English lords let peasants graze their sheep on their land but punished them for removing any droppings (Driver et al. 1999).

German alchemist Hennig Brand is the earliest known to discover the pure form of phosphorus in 1669.[1] He did so somewhat accidentally, during his hunt for the

---

[1] There may have been earlier discoveries in ancient Rome. Ashley, K., Cordell, D. & Mavinic, D. (2011) A brief history of phosphorus: From the philosopher's stone to nutrient recovery and reuse. *Chemosphere*, 84, 737–746.

Fig. 15.1 Aboriginals making fire (Mützel 1857, PD (PD: This image has been released to the public domain))

philosopher's stone—a stone that would turn base metals into gold. Brand was distilling large quantities of urine, extracting phosphorus, and then cooling it to turn into a solid (see Fig. 15.2). Although the solid form did not achieve the goal of turning things to gold, it did glow in the dark. He did not reveal his discovery until 1675, when he and colleague Daniel Kraft became famous as they presented their new form of light. Phosphorus was not recognized as an element until a century later when Antoine Lavoisier, the founder of modern chemistry, finally recognized it as such (Ashley et al. 2011).

From the 1700s to the early 1800s, phosphorus was used widely for medicinal purposes. Johann Linck was the first to sell phosphorus as a medicine in 1710. He suggested his pills could cure colic, asthmatic fevers, tetanus, apoplexy, and gout. His pills allegedly contained 200 mg of phosphorus. However, doses as low as 1 mg/kg can be lethal, so it seems his claims were somewhat exaggerated. In the mid-1700s, Dr. Alphonse Leroy prescribed phosphorus as a sexual enhancement agent on the basis of self-experimentation (Emsley 2002).

The only medicinal purpose for which phosphorus was actually effective was for abortion, but at high risk to the mother. It is only a little more toxic to a foetus than the mother, but in the late nineteenth century, it was frequently used for this purpose. Women would scrape the heads off matches to access phosphorus. Over 1400

Fig. 15.2 The Alchemist in search of the philosopher's stone (Wright 1771, PD (PD: This image has been released to the public domain))

events of poisoning were recorded in Sweden between 1851 and 1903. Only 10 mothers survived (Shorter 1991).

It was not until 1840 that people first understood the chemistry behind how dead and decaying matter created new life (Liebig 1840; Ashley et al. 2011). Even once the chemistry was understood, human activity did not substantially alter the phosphorous cycle until much later. Famine and soil degradation led to the use of external sources of phosphorus. Phosphorus was removed from the soil with crops, but it was replaced with organic phosphorus such as crushed or dissolved bones, guano (bird droppings), human waste, crop residue, and manure (Emsley 2002) (see Fig. 15.3).

After World War II, the use of mineral phosphorus from phosphate rocks grew exponentially (Science Communication Unit 2013). A combination of the sanitation revolution, which shifted the deposit of human excrement from land to water, and the green revolution, the discovery that synthetic fertilizers made from phosphorus and nitrogen could significantly improve crop yield, transformed agricultural practices.

As with many changes, the transition was not slow and linear, but abrupt and system changing. From 1950 to 2000, mineral-fertilizer use grew sixfold (Science Communication Unit 2013). It became feasible to separate the production and consumption of crops over longer distances due to the ability to transport synthetic fertilizers. Arable farming could now be separated from livestock farming.

Fig. 15.3 Bison skulls were dissolved for phosphorus (Unkown 1892, CC PD 3.0 (CC BY-SA 3.0: Creative Commons licence allows reuse with appropriate credit))

Manure went from being a valuable and important resource to a waste product (Schröder et al. 2010).

15.3.1   The (Human) Phosphorous Cycle

The seeming efficiencies of synthetic phosphorous fertilizer came with many problems. Although crop yields increased, less care was taken with recycling waste products back into society. The previously circular, closed-loop system had become linear. Phosphorus was (and is still) extracted, used, and disposed of in a mostly linear system worldwide.

The life cycle of mined phosphorus is not only environmentally damaging but also hugely inefficient. Of the phosphorus mined for fertilizer, it is estimated that only one fifth is present by the time it is consumed (Cordell et al. 2009). The process, often termed "mine to fork", has losses and environmental impacts at every stage from the extraction and primary processing to the processing, fertiliser application, and harvesting and postharvesting:

15.3.1.1   Extraction

Before phosphorus can be extracted, the mine site needs to be prepared. Impacts at this stage include the clearing of vegetation, topsoil removal, and the removal of overlying rock. These changes can lead to changes in surface and underground water flow patterns, topography, habitats, and biodiversity.

Both the mine preparation and the extraction of phosphorus are very energy-intensive processes and therefore incur high levels of associated CO_2 emissions. Other impacts include substantial water consumption and soil erosion. Approximately 18% of the phosphorus mined is lost through inefficiencies at this stage (Prud'homme 2010).

15.3.1.2   Primary Processing

Phosphate rock is often associated with contaminants such as cadmium which is toxic and uranium which is radioactive. Not only phosphorus but also these contaminants can be lost to the environment during primary processing, i.e. beneficiation and cleaning. Average losses at this stage are 16% (Schröder et al. 2010).

15.3.1.3   Processing

Approximately 14.9 MtP/year is processed from phosphate rock into phosphate products globally. This can be done using acid (to develop fertilizer) or heat (to develop industrial phosphorus and feed phosphates). Phosphogypsum, a by-product of processing phosphate into fertilizer using sulphuric acid, is one of the more harmful waste products in the human-phosphorous cycle. It is usually mixed with water to make a slurry and then deposited on land to allow the solids to settle out—a process known as wet stacking. The concern is that radioactive material in the slurry could leach into groundwater. Some phosphoric acid plants do not even take these precautions with wet stacking and simply release the slurry into freshwater bodies and even into the oceans (Wissa 2003). Approximately 4–5 tonnes of phosphogypsum are generated per tonne of phosphoric acid. Phosphorus losses at the processing stage range from approximately 5% for acid processing to 10% for heat processing (Prud'homme 2010).

15.3.1.4   Fertilizer Application

Only a third of the phosphorus in fertilizers is absorbed by plants. The rest accumulates in soil (in which case it is not really lost as it is still available to future plants), is washed away by rainwater, or is blown by the wind in soil or water particles (Science Communication Unit 2013). There are other minor losses at this stage from pests and diseases, but this phosphorus is usually redeposited and available for plant use. The impacts of phosphorus lost to water bodies are the basis for the inclusion of phosphorus in the Planetary Boundaries framework (Rockström et al. 2009b). Phosphorus promotes algal growth which can lead to anoxic events in water bodies that can wipe out the entire marine or freshwater ecosystems.

15.3.1.5   Harvest and Postharvest

During harvesting, crop residues such as husks account for further phosphorous losses. Some crop residues are left on site where they are generally returned to the soil. In this case, the phosphorus is not lost as it can be reused in the next crop rotation.

Food waste, excreta, and animal feed losses may account for as much as 30% additional losses during the postharvest stage (Kantor et al. 1997).

At each stage, there are also minor losses such as spillages, spoilage, theft, or being lost in storage or transport (Isherwood 2000).

As discussed above, some of the apparent losses are actually an accumulation in the soil. Little was known or considered about the different phosphorous needs of different soil types until recently, and so in some cases, far more phosphorus has been applied than needed. In the Netherlands, there is enough phosphorus in the soil to supply the country with phosphorus for the next 40 years (Wilt and Schuiling n.d.).

Almost 100% of phosphorus consumed as food is excreted in urine and faeces. Approximately 70% is in urine, and the remaining 30% in faeces to a total of 3 Mt/year. Less than 10% of this phosphorus is reused. Of this, most of this is used indirectly as untreated or treated wastewater, some as sludge, and some from ash (from incinerated sludge). A small amount is also used directly via composting toilets and direct defecation. The remaining 90% is discharged to water or land (Cordell et al. 2009).

Past practices of phosphorous use and management were sustainable as humans tapped into and expanded the natural phosphorous cycle without fundamentally altering it. In contrast, current use is unsustainable. We now take substantial amounts of phosphorus from the inorganic cycle, use it once, and release it into the organic cycle at rates that the organic cycle cannot process.

15.3.2   Phosphorus: A Nonrenewable Resource

Mineral phosphorus found in phosphate rock is a nonrenewable resource as the rate of replenishment of phosphate in rock, which occurs over millions of years, is of a different order of magnitude than human activity. There are 4×10^{15} tonnes of phosphorus in the Earth's crust. Humans currently consume approximately 3×10^6 tonnes of phosphorus per year (Schröder et al. 2010). The problem is that very little of the phosphorus in the crust is accessible. Much of the phosphorus that is accessible is either in such low concentrations that extraction is not economically viable or there is too much contamination by other substances (MEA 2005).

The exact amount of phosphate rock reserves are difficult to determine. Reserves are defined as phosphate rock that is accessible using existing technology and is economically viable. The most recent estimates by the International Fertilizer Development Centre (IFDC) (2010) are that 60,000 billion tonnes of phosphate rock reserves remain. This is substantially higher than the previous US Geological Survey estimate of 16,000 (Science Communication Unit 2013). As technology improves or as demand for phosphorus increases, the amount of phosphate rock that is deemed accessible or economically viable is likely to increase.

The question of whether we are facing an imminent supply shortage of phosphorus is debated in the literature. Estimates of the amount of high-quality phosphate rock remaining range from only a few decades worth to a few hundred years (Schröder et al. 2010). Van Vuuren et al. (2010) in Science Communication Unit (2013) assessed phosphorous levels under different scenarios around agriculture, household, and sewage systems and found that there were no signs of near-term depletion. However, they did find that longer-term, low-cost, high-grade resources would be in short supply. This is consistent with other studies (e.g. Schröder et al. 2010; Science Communication Unit 2013).

Not only is phosphate rock nonrenewable; it is also very unevenly distributed across the globe. Almost 75% of the known reserves are located in Morocco and Western Sahara. A further 20% is located in China, Algeria, Syria, Jordan, South Africa, the United States, Russia, Peru, Australia, and Saudi Arabia. Ninety-five percent of the reserves are controlled by only ten countries. Of these countries, most of the exports are from Morocco and Jordan. China, the United States, and South Africa use their reserves in-house (Science Communication Unit 2013).

Moreover, the ownership structures of the mines and supply chains suggest market volatility is likely (Elser and Bennett 2011). The mines in Morocco are state owned. Given the large proportion of global phosphorus that is located in Morocco, this puts the Moroccan government in a position of power to control the market price. Additionally, large parts of the supply chains globally (i.e. the mining, processing, and fertilizer production) are operated by a single firm. This sort of vertical integration has been shown to increase the likelihood of monopolization (De Ridder et al. 2012).

Most countries are heavily reliant on imports of phosphorus, and there has already been evidence of market volatility. In 2008, there was an 800% price spike for phosphorus. After this spike, China imposed a 135% export tariff on phosphates (Fertiliser Week 2008). The Arab Spring in Tunisia led to a 40% drop in exports of phosphorus (De Ridder et al. 2012).

15.4   An Indicator for Phosphorus

It is not the potential scarcity of phosphorus, but the environmental impacts of its use that have led it to be included as a Planetary Boundary limit. When excessive levels of phosphorus make their way into water bodies, this can lead to algal blooms or eutrophication (Schröder et al. 2010). The intense blooms block sunlight from entering the water below which reduces the amount of oxygen dissolved in that water, thus creating anoxic conditions or "dead zones". Originally the sea had very little oxygen, and only single-celled organisms with low oxygen needs were able to survive. As oxygen levels increased, so did aquatic life. However, there is evidence that there were at least partial returns to oxygen-free oceans in our history. Indeed, it is thought that past phosphorus inflow into the oceans may have been the primary cause of global scale ocean anoxic events which lead to mass extinctions of marine life (Handoh and Lenton 2003). There are currently more than 400 costal dead zones in the oceans from phosphorus with large dead zones in the Gulf of Mexico, the Baltic Sea, and the Atlantic off West Africa. The environmental impacts of algal blooms continue after the death of the algae as this releases toxic compounds which can also kill fish in surrounding waters (Correll 1998).

There are requirements to treat wastewater to prevent this from occurring in many developed countries with regulations such as the Urban Wastewater Treatment Directive (EEC 1991). However, the success of these regulations varies. In Europe the amount of water treated ranges from 4% to more than 97% (OECD 2004).

The primary Planetary Boundary limit for phosphorus is a flow of no more than 11 TgP/year from freshwater systems to the ocean. This limit is set at a point where the risk of a global anoxic ocean event is considered low. There is also a secondary limit of a flow of no more than 6.2 TgP/year from fertilizers to erodible soils (Rockström et al. 2009a).

The PB control variable is a Pressure indicator.[2] However, in this instance, the flow is describing the movement of a substance between environments. It is not describing a flow from human activity. This means that it is difficult to compare this control variable directly to human activity, one of the criteria for selecting PQ indicators (see Chap. 7). However, the maximum flow of phosphorus from freshwater systems to the ocean is the amount of phosphorus released to the environment by human activity. This is also a Pressure indicator and one which can be easily related to any scale of human activity. As such, this is the indicator selected for the PQ for phosphorus.

---

[2] Environmental indicators can be classified as Drivers, Pressures, States, and Impacts under the EU DPSIR framework. A Pressure is an indicator that describes an environmental flow and is also the category of indicator used for the Planetary Quotas. See Chap. 5 for further details.

15.5   The Limit

Over a long timeframe, it can be assumed that almost all phosphorus released to the environment by humans will end up in the oceans. As such, the PQ limit for phosphorus should be the same as the PB limit. This means that the PQ for phosphorus is *net phosphorus released to the environment* ≤ 11 Tg/year.

15.6   Discussion

The current rate of phosphorus flowing from freshwater systems to the ocean is approximately 22 Tg/year. This means that the PQ for phosphorus is currently being exceeded.

Table 15.1 lists examples of activities which could occur at different scales and across different areas of the community in order to manage human use of phosphorus.

15.7   Conclusions

Phosphorus is a nonrenewable substance that is critical to life on Earth. However, human use of mined phosphorus as fertilizer is having severe downstream impacts. This nutrient can stimulate unnatural levels of algal growth that cause anoxic events wiping out the entire ecosystems.

Table 15.1 Examples of different scales of activity which have or could contribute to achieving the PQ for phosphorus

Achieving the Planetary Quota for phosphorus			
	Community	**Government**	**Business**
Large	Develop a global organization dedicated to the measurement and management of phosphorous use	Develop legislation around maximum phosphorous applications and phosphorous management practices	Develop phosphorous recycling techniques to reduce demand on raw phosphorous supplies and supply of waste phosphorus to waterways
Medium	Campaign for greater awareness about the environmental impacts of phosphorus	Trial regulations on phosphorus requiring recycling	Build a business around phosphorous recycling
Small	Community-based composting and food production systems (e.g. permaculture)	Facilitate community-based systems	Invest in small-scale recycling and food systems
Individual	Eat an organic and plant-based diet		

The Planetary Quota for phosphorus is net phosphorus released to the environment ≤ 11 Tg/year. This can be compared to the amount of phosphorus released during any scale of human activity, the phosphorus footprint.

References

Ashley K, Cordell D, Mavinic D (2011) A brief history of phosphorus: from the philosopher's stone to nutrient recovery and reuse. Chemosphere 84:737–746
Cordell D (2001) Improving carrying capacity determination: material flux analysis of phosphorus through sustainable aboriginal communities. University of New South Wales (UNSW), Sydney
Cordell D, Drangert J-O, White S (2009) The story of phosphorus: Global food security and food for thought. Glob Environ Chang 19:292–305
Correll DL (1998) The role of phosphorus in the eutrophication of receiving waters: a review. J Environ Qual 27:261–266
De Ridder M, De Jong S, Polchar J, Lingemann S (2012) Risks and opportunities in the Global phosphate rock market. The Hague Centre for Strategic Studies (HCSS), The Hague
Driver J, Lijmbach D, Steen I (1999) Why recover phosphorus for recycling, and how? Environ Technol 20:651–662
EEC (1991) Council directive concerning urban waste water treatment. The Council of the European Communities, EUR-Lex
Elser J, Bennett E (2011) A broken biogeochemical cycle. Nature 478:29
Emsley J (2002) The sordid tale of murder, fire and phosphorus—the 13th element. Wiley, Chichester
Fertiliser Week (2008) Industry ponders the impact of China's trade policy. Thursday market reports. British Sulphur Consultants, CRU
Flannery T (1994) Future Eaters: an ecological history of australasian lands and people, NSW, Australia. Reed Books, Chatswood
Föllmi KB (1996) The phosphorus cycle, phosphogenesis and marine phosphate-rich deposits. Earth-Sci Rev 40:55–124
Handoh IC, Lenton TM (2003) Periodic mid-Cretaceous oceanic anoxic events linked by oscillations of the phosphorus and oxygen biogeochemical cycles. Glob Biogeochem Cycle 17
IFDC (2010) 35 years of accomplishments: building for the future. International Fertiliser Development Centre, Muscle Shoals
Isherwood K (2000) Mineral fertilizer distribution and the environment. UNEP, Nairobi
Kantor LS, Lipton K, Manchester A, Oliveira V (1997) Estimating and addressing America's food losses. Food Rev 20:2–12
Liebig JFV (1840) Die organische Chemie in ihrer Anwendung auf Agricultur und Physiologie (Organic chemistry in its applications to agriculture and physiology). F. Vieweg, Braunschweig
Liu Y, Villalba G, Ayres RU, Schroder H (2008) Global phosphorus flows and environmental impacts from a consumption perspective. J Ind Ecol 12:229–247
MEA (2005) Ecosystems and human well-being: biodiversity synthesis. World Resources Institute, Washington, DC
Mützel G (1857) Aboriginal fire making.jpg. Wikimedia
OECD (2004) The OECD environmental strategy: progress in managing water resources. OECD observer. Organisation for Economic Cooperation (OECD), Paris
Prud'homme M (2010) World phosphate rock flows, losses, and uses. The phosphates 2010 conference and exhibition, 22–24 March 2010. Brussels, Belgium
Rockström J, Steffen W, Noone K, Persson Å, Chapin FS, Lambin EF, Lenton TM, Scheffer M, Folke C, Schellnhuber HJ, Nykvist B, DE Wit CA, Hughes T, Van Der Leeuw S, Rodhe H, Sörlin S, Snyder PK, Costanza R, Svedin U, Falkenmark M, Karlberg L, Corell RW, Fabry VJ,

Hansen J, Walker B, Liverman D, Richardson K, Crutzen P, Foley JA (2009a) A safe operating space for humanity. Nature 461:472–475

Rockström J, Steffen W, Noone K, Persson A, Chapin FS III, Lambin E, Lenton TM, Scheffer M, Folke C, Schellnhuber HJ, NykvisT B, De Wit CA, Hughes T, Van Der Leeuw S, Rodhe H, Sörlin S, Snyder PK, Costanza R, Svedin U, Falkenmark M, Karlberg L, Corell RW, Fabry VJ, Hansen J, Walker B, Liverman D, Richardson K, Crutzen P, Foley J (2009b) Planetary boundaries: exploring the safe operating space for humanity. Ecol Soc 14

Schlesinger WH, Bernhardt ES (2013) Biogeochemistry: an analysis of global change, 3rd edn. Elsevier, Amsterdam

Schröder JJ, Cordell D, Smit AL, Rosemarin A (2010) Sustainable use of phosphorus. Plant Research International—Wageningen University and Research Centre, Wageningen

Science Communication Unit (2013) Science for environment policy in-depth report: sustainable phosphorus use. Report produced for the European Commission DG Environment, Bristol

Shorter E (1991) Women's bodies: a social history of women's encounter with health, ill-health, and medicine. Routledge, London

Unkown (1892) Bison skull pile-restored.jpg. Wikimedia

Wilt J, Schuiling O (n.d.) Urgentie en opties van onderzoek en beleid. Universiteit van Utrecht, Utrecht

Wissa A (2003) Phosphogypsum disposal and the environment. Ardaman & Associates, Inc, Orlando, FL

Wright J (1771) The Alchemist in search of the philosophers stone. Wikipedia

Chapter 16
The Biodiversity Quota

We should preserve every scrap of biodiversity as priceless
while we learn to use it and come to understand what it means
to humanity
E.O. Wilson

Abstract Recent human activity has had more severe impacts on species loss than any other period in human history. Despite efforts to manage this, the impacts are continuing to increase. Biodiversity is very important to the Earth-system function and to humanity directly because of the ecosystem services it provides.

It is extremely difficult to link biosphere health to human activity as there are so many different ways that human activity can be damaging to the biosphere. Human impacts on biodiversity through climate change and pollution are addressed through Planetary Quotas for carbon dioxide, MeNO, aerosols, Montreal gases, and forest-land. Habitat destruction and segregation is one of the greatest human drivers of biodiversity loss. A new proxy indicator has been developed by the UNEP to link land use to pressures on biosphere integrity, "percentage disappearing species". It is an estimation of species extinctions caused through land-use change. This indicator is the best proxy indicator available with which to assess the effects on human activity on global extinction rate.

The Planetary Quota for biodiversity is percentage disappeared fraction of species $\leq 1 \times 10^{-4}$/y.

16.1 Introduction

The extinction of species is a natural process. Almost every species that has ever lived on Earth is already extinct. Species extinction will occur with or without human intervention. However, as a result of human activity, the current rate of biodiversity loss is 10–100 times greater than estimated natural rates (Secretariat of the CBD 2001).

© Springer Nature Singapore Pte Ltd. 2020
K. Meyer, P. Newman, *Planetary Accounting*,
https://doi.org/10.1007/978-981-15-1443-2_16

Biodiversity can be defined as the extent of variability in plant and animal species. Biodiversity is critical to the functioning of the Earth system. Different species play different roles in an ecosystem, and the loss of one species can sometimes mean the collapse of an entire ecosystem. The bee, for example, is responsible for the pollination of many different plant species. Without the bee, many of these plants risk dying out. Species with particularly important roles within their ecosystems are referred to as keystone species.

Some believe that managing human impacts on biodiversity is a far more challenging task than reducing emissions as it is so difficult to attribute species threats to human activity (Moran et al. 2016; Vačkář 2012). Consumers are often very disconnected from the impacts of their consumption on biodiversity health. In one study, as much as 44% of the threats to species from net exporting countries were found to occur outside their national boundaries (Lenzen et al. 2012a).

Of the nine Planetary Boundaries, biosphere integrity is one of the limits we have transgressed the most. It is also the most interconnected PB. The PB indicator for biodiversity loss—extinction rate—is associated with every Planetary Quota (see Chap. 7, Fig. 7.4). This high level of interconnectivity has led the PB for biosphere integrity to be considered as one of two core PBs (climate change being the other). Both are intrinsically connected with almost every other PB.

This chapter shows how the PQs can be used for connecting human activity to unwanted biodiversity-loss outcomes. It begins with some background about species extinctions—the main drivers – and the past mass extinction events. This is followed by an overview of how biodiversity health and in particular how the relationship between this and human activity has been measured. The case is then made for the proxy indicator and proposed PQ limit. The chapter concludes by putting this limit into context and providing examples of the sorts of activity that might be needed to address this PQ.

16.2 Background

The biosphere is the part of Earth where there is life—a thin envelope around Earth's surface. Most organisms depend directly or indirectly on sunlight, so life is predominantly located where sunlight has access—i.e. the surface, the atmosphere, the upper layers of the oceans and lakes, and the top layer of soil. There is life deeper in the oceans and in Earth's crust—bacteria live almost everywhere. A new project is underway to drill into Earth's mantle to determine, among other things, whether there are bacteria or any forms of life in Earth's mantle. Biodiversity is highest at the equator and reduces towards the poles, with moist forests in the tropics providing the greatest species richness (Secretariat of the CBD 2001).

Humans rely on biodiversity to fill many functions, often referred to as ecosystem services. The second report by the Convention on Biological Diversity listed 24 ecosystem services performed by species in the biosphere (Secretariat of the CBD 2006). Ecosystem services include:

- Ecosystem services such as the provision of food from plants and animals.
- The provision of resources such as timber and bioenergy and biotechnology (the use of living organisms such as yeast for bread and beer).
- Regulating services including as water filtration, decomposition of organic waste, and climate regulation.
- Supporting services such as nutrient cycling (e.g., the nitrogen and phosphorous cycles) and photosynthesis.

Genetic diversity is the diversity of the genetic makeup of a single species. Genetic diversity is important. It helps species to be flexible in harsh conditions such as climate change, storms, or widespread outbreaks of pests (Secretariat of the CBD 2001). Genetic material taken from wild species is used to improve crops and develop drugs and as raw materials for use in products and services.

Diversity of species and within species is necessary for resilient ecosystems and for a resilient biosphere. There are also moral, ethical, cultural, aesthetic, and scientific reasons to conserve biodiversity. However, it is biodiversity's role in the regulation of the Earth system that is the basis for the Planetary Boundary for biosphere integrity.

Fossil records lead us to expect approximately one species to become extinct every 400 years (birds) and 800 years (mammals). Over the last 400 years, it is estimated that we have been losing 20–25 species every 100 years. This is 100–200 times faster than the base level. Species extinction can only be measured through negative evidence, i.e. a lack of species. This means that monitoring of extinction rates is very limited, and it is hard to say how many species have gone extinct over a period or to predict likely future extinctions. It is possible that the extinction rates are substantially higher than those estimated. To accumulate negative evidence, there is substantial lag as until sufficient time has passed, it is hard to say whether there are few or none of a particular species (Mace et al. 2014).

Most known plant and animal extinctions have been on islands, and most continental extinctions have been on freshwater organisms. There are few extinctions recorded in continental rainforests. However, the monitoring of this is extremely difficult. The rate of *known* biodiversity loss in the oceans is much less than in any other type of ecosystem. This may be due to a lack of knowledge of the extinctions but is also likely to be related to the size of the oceans and the fact that people do not live in them permanently.

Human activity is generally thought to negatively impact biodiversity. In the case of genetic biodiversity, however, human activity can and has both reduced and increased diversity (Secretariat of the CBD 2001). Humans have been indirectly but purposefully manipulating biodiversity for more than 10,000 years. This has led to the current high diversity of domesticated crops and livestock. The manipulation has been indirect because the focus was on preserving or developing features rather than genes themselves, for example, efforts to increase pest resistance or milk yield. Humans also alter genetic diversity intentionally through genetic engineering. This is the introduction of a section of DNA from one organism into another, where it would not naturally occur, to produce a genetically modified organism (GMO) with favourable properties.

Although human activity does influence biodiversity in both directions, it is overwhelmingly harmful. In the last 50 years, human activity has had more impacts on biodiversity than in any other period in human history (MEA 2005). Despite increased efforts to reduce impacts, it has been estimated that the impacts would continue to worsen until at least 2020 (Secretariat of the CBD 2014; Tittensor et al. 2014).

Between 2000 and 2012, 0.5 Mkm^3 of tropical rainforests—the most biodiversity-rich habitats—were destroyed (Hansen et al. 2013). Human activity has led to increases in global average temperatures that have shifted and eliminated habitats. In the late 1990s, marine capture fisheries was almost 90 million tonnes. It is estimated that 93% of fish stocks are either overfished (33%) or maximally sustainably fished (60%) (FAO 2018). Other impacts on marine ecosystems include waste disposal, recreation, coastal stabilization, and transportation especially shipping that transfers feral species in their ballast tanks or on their hulls. Chemical pollution and eutrophication is widespread and is now plastic waste-based. Fishery operations can also destroy the seabed and affect population levels of nontarget species. Commercial bottom fishing disturbs seafloor organisms and the seabed impacting both habitats and species.

Biodiversity loss can cause permanent changes to the planet (Rockström et al. 2009). Changes to the planet can also cause permanent and extensive biodiversity loss. A global mass extinction is defined as a period where more than 75% of species become extinct. This has happened five times in known history as set out below.

Ordovician-Silurian Extinction.

The first known global mass extinction occurred approximately 444 million years ago, at the end of the Ordovician period. Approximately 85% of species were lost. There are thought to have been two drivers for the extinctions—a brief but severe Ice Age and falling sea levels. Both drivers may have been the result of the uplifting of the Appalachians—a mountain range. The mountains were previously unexposed silicate rock. Once exposed, the silicate rapidly absorbed CO_2 out of the atmosphere leading to rapid global cooling. The changing land mass and glaciation could have driven the sea level fall (Harper et al. 2014).

Late Devonian Extinction.

The second mass extinction is thought to have occurred approximately 375 million years ago during the Devonian period. Seventy-five percent of species were lost. It is thought that new land plants with deep roots may have caused the extinctions. The roots stirred up the Earth, releasing nutrients into the ocean. This may have triggered algal blooms that sucked the oxygen out of the water, suffocating marine species.

Permian-Triassic Extinction.

The third mass extinction event occurred at the end of the Permian period—approximately 251 million years ago. This was the worst extinction event, known as "the great dying", with a species toll of 96%. Scientists suggest that this event set life back by 300 million years. The event is believed to have been set in motion by a volcanic eruption near Siberia. Huge amounts of CO_2 were released into the atmosphere. In response, bacteria released vast quantities of methane. The greenhouse gases warmed the atmosphere, while the high levels of CO_2 in the atmosphere caused the oceans to become acidic and stagnated.

Triassic-Jurassic Extinction.
Two hundred million years ago, at the end of the Triassic period saw a loss of 80% of species. There is no clear cause for this mass extinction.
Cretaceous-Paleogene Extinction.
Sixty-six million years ago, was the most recent and best known of the mass extinctions—the demise of the dinosaurs. This was caused by an asteroid hitting the Earth in the Mexican area.

The current rate of known species loss is of an order of magnitude that is approaching that of these global major extinction events (Chapin et al. 2000). This is due to clearing forests, spreading of feral animals and weeds, overexploitation of ecosystems and pollution, including climate change. Jonathan Watts in *The Guardian* (Watts 2019) suggests that "This year, the world's leading scientists *warned that human civilisation was in jeopardy* because forest clearance, land-use shifts, pollution and climate change had put a million species at risk of extinction".

16.2.1 Biodiversity Management

There have been efforts to preserve biodiversity on a global scale since the 1980s. In 1988 the United Nations Environment Programme (UNEP) convened the Ad Hoc Working Group of Experts on Biological Diversity. The following year UNEP started the Ad Hoc Working Group of Technical and Legal Experts later known as the Intergovernmental Negotiating Committee, who were tasked with the development of a legal framework for conserving biodiversity.

Their work became the Convention on Biological Diversity (CBD) which was adopted in Nairobi in 1992. It entered into force in December 1993 (CBD 2018).

The CBD is a legally binding global treaty. It covers:

- Conservation of biodiversity.
- Sustainable use of its components.
- Fair and equitable sharing of benefits rising from the use of genetic resources.

Participation in the convention is nearly universal. In 2001 the CBD produced its first periodical report on the state of global biodiversity—Global Biodiversity Outlook 1. There have been four such reports, the most recent released in 2014.

In 2002 the conference of the parties adopted a Strategic Plan with the mission "to achieve, by 2010, a significant reduction of the current rate of biodiversity loss at the global, regional, and national level, as a contribution to poverty alleviation and to benefit all life on Earth". It was endorsed by the Heads of State and Government at the World Summit on Sustainable Development in Johannesburg. At the 2005 World Summit of the United Nations (UN), world leaders reiterated their commitment to the 2010 targets (Secretariat of the CBD 2006).

The COP established supporting goals and targets and identified indicators for evaluating biodiversity status and trends (Secretariat of the CBD 2006). The targets were not met (Butchart et al. 2010). It is thought that the failure to meet the targets is because efforts were not aimed at the underlying causes of biodiversity loss (Secretariat of the CBD 2010). The responses were predominantly focussed on the direct pressures and on the state of biodiversity (Secretariat of the CBD 2010).

The UN General Assembly designated the period 2011–2020 as the UN Decade on Biodiversity. The CBD developed a Strategic Plan for Biodiversity from 2011 to 2020 with a 2050 vision of the end to biodiversity loss and a move to sustainable use of ecosystems. This included five strategic goals and 20 Aichi targets (Secretariat of the CBD 2014).

Box 16.1 Poly-Scalar Management and Biodiversity

There is a lot of reference to poly-scalar management approaches within the CBD reports:

- In the second report from the CBD (2006), the authors highlighted the need for action at all levels, i.e. a poly-scalar approach (see Chap. 4).
- In the third report, the idea was repeated. The report suggests: "biodiversity loss could be slowed and even stopped if Governments and society took coordinated action at a number of levels".
- In the 2011–2020 strategic plan, the point was made again, this time with more strength: "the basis of the Stategic Plan is that biodiversity loss can only be effectively addressed with simultaneous and coordinated action at a number of levels, each of which is essential to achieve a lasting impact and to set us on a sustainable path to keep human societies within the limits of the planet's biological resources".
- In the fourth report, the CBD repeated the need to address underlying drivers through change at all levels (Secretariat of the CBD 2014). However, within the same report is the statement: "It is therefore an appropriate opportunity to review progress towards the goals of the Strategic Plan, and to assess what further action governments may need to take to achieve the targets they collectively committed to in 2010".

It is promising to see that the science of governance and change is becoming integrated with efforts to manage the environment.

16.2.2   Measuring Biodiversity Health

There is currently no consensus on best ways to measure biodiversity health (Moran et al. 2016). It is difficult to accurately assess the health of biodiversity in a given ecosystem. It is even harder to relate threats to biodiversity to human activity.

A biodiversity footprint has been proposed—as the number of species threatened due to land conversion, land-use changes, unsustainable use of natural resources, overexploitation of marine ecosystems, and invasive alien species (Lenzen et al. 2012a; Cucek et al. 2012). However, this indicator is still an Impact indicator. The measure does not relate to the drivers, but to the outcome—the "number of species threatened".

Some authors have attempted to connect biodiversity health to human activity. Asafu-Adjaye (2003) found that, while economic growth has an adverse effect on biodiversity, the composition of the economic output is important. There is a theory called the Kuznets hypothesis, which is that environmental impacts will increase with higher affluence to a certain point but then reduce as affluence grows beyond this point and people have different priorities beyond basic survival. The hypothesis does appear to work with issues like air pollution and climate change mitigation especially when the extra wealth is associated with greater urbanization (ADB 2012). Dietz and Adger (2003) tested this hypothesis for biodiversity loss but found that the curve did not exist. Another study concurred that the curve was not present for most taxonomic groups (Naidoo and Adamowicz 2001). However, these authors found that the number of threatened birds did drop as gross national product increased. As a general rule, the number of threatened species increases with population and with gross national product, though this may change as so much of the threatened rainforests are in the developing world.

A new, binary certification has recently been developed to account for whether or not a product or sector exerts pressure on endangered species (Moran et al. 2016). This is a tidy parallel to the PB indicator—extinction rate. Such a system could potentially be used to certify products as biodiversity certified to account for the biodiversity pressures not included through the PQs. Moran et al. use a 1 or 0 for each sector—either they do exert pressure or they do not. They do not attempt to measure the amount of pressure. They use four case studies to identify that environmental impacts can be traced to products through supply chain via input-output analysis. This information could also be included in a "Planetary Facts" product labelling system (see Chap. 17). Environmental labelling has been widely used to indicate impacts to consumers (Moran et al. 2016). This binary certification system would address the difficulty of communicating whether or not a product or service was impacting species extinctions. However, this indicator cannot be scaled, allocated, or easily connected to any scale of activity. It is thus not suitable as the PQ indicator.

The Leontief calculus (1986), a model for the economics of a country or region, has also been used to connect final consumers with upstream biodiversity impacts. The basis of the model is to determine the quantity of a primary resource, for example, coal, that would be needed to supply $1 of demand for every consumer in the country or region. Lenzen et al. (2012b) have used this concept to determine the number of species endangered by the development of a product by determining the impacts per dollar of product for a given year. The problem with this method is that it relies on input-output calculations which are typically based on national level imports and exports. Very few countries have input-output tables at a subnational level where most biodiversity impacts occur.

16.3   The Indicator

There is both a global and a regional Planetary Boundary for biosphere integrity (Steffen et al. 2015):

- Global: ≤10 extinctions per million species per year (E/MSY) (with an aspirational goal of ≤1 E/MSY).
- Regional: Biodiversity Intactness Index of ≥90% (with uncertainty range 90–30%).

The global PB is based on the extinction rate over the past several million years (Steffen et al. 2015). In both the first and second PB articles, the authors expressed a high level of uncertainty around both the control variable and the threshold for this PB. The proposed control variable has been criticized for being difficult to assess accurately or in a timely manner and, importantly, for being an unsuitable metric to apply to different scales (Mace et al. 2014).

The drivers of species extinctions, or biodiversity loss, are complex and not completely understood (Secretariat of the CBD 2014; Vačkář 2012). However, most of the literature agrees that the five primary anthropogenic threats contributing to biodiversity loss are:

(a) Climate change—shifting habitat to an extent that it is no longer suitable for the threatened species.
(b) Pollution that affects the health of species.
(c) Overexploitation of species, especially due to fishing and hunting but also overuse of ecosystem services leading to aforementioned habitat loss.
(d) Spread of invasive species or genes outcompeting endogenous species.
(e) Habitat loss; fragmentation or change, especially due to agriculture; large-scale forestry; and human infrastructure.

(Galli et al. 2014; Rockström et al. 2009; Cucek et al. 2012; Secretariat of the CBD 2014; MEA 2005).

There is some debate as to the relative impacts of these threats with respect to one another. In one study comparing threats to species in the United States, land use was found to be the greatest threat, affecting 85% of species. Invasive alien species was found to be the second highest threat for most species, affecting 49%. However, pollution was the second highest threat for aquatic species (Wilcove et al. 1998).

Lenzen et al. (2012a) have estimated that 30% of global species threats, excluding threats from invasive alien species, can be attributed to international trade. The same study showed that for the net importing countries studied, as much as 44% of their biodiversity footprint occurred outside of their national boundaries. This means that net exporting countries have very high biodiversity tolls. Approximately 35% of the threats to biodiversity in net exporting countries were related to production for export.

Others believe the most significant impact is change in land use such as conversion of ecosystems into agricultural and urban areas, changes to frequency, duration or magnitude of wildfires, and introduction of new species (Secretariat of the CBD

Table 16.1 The key threats to biodiversity identified for different ecosystems in GBO1 under the five threat categories identified in the literature

Ecosystem	Habitat loss/ change	Pollution	Climate change	Overexploitation	Introduced species
Inland water ecosystems	• Alteration and destruction of habitat through water drainage, canalization, and flood control • Construction of dams and reservoirs • Sedimentation	• Sedimentation • Pollution: ∘ Eutrophication ∘ Acid deposition, salinization ∘ Heavy metals			• Introduced species
Forests	• Conversion to cropland and plantations • Conversion to urban or industrial land • Fragmentation • Changing fire regimes	• Pollutants, including acid rain	• Changing fire regimes • Climate change	• Logging • Extraction of non-timber forest products • Fuelwood extraction • Hunting • Unsustainable shifting cultivation	• Invasive alien species
Drylands	• Conversion to cropland	• Chemical inputs – artificial enrichment	• Changing fire regimes • Climate change	• Water use • Depletion of groundwater resources • Harvest of wood for fuel • Overharvest of wild species	• Introduced herbivores, particularly livestock • Introduction of pathogens • Introduction of non-native plants

2010; MEA 2005; Fahrig 2001; Groombridge 1992; Bibby 1994; Ehrlich 1994; Thomas et al. 1994; Wilcove et al. 1998).

The first global report on biodiversity (Secretariat of the CBD 2001) differentiates specific threats to species based on ecosystem type. The threats listed in this report can each be classified under one of the five key threats identified above. Table 16.1 shows that most of the key threats are applicable to most types of ecosystem.

The second global report on biodiversity (Secretariat of the CBD 2006) shows the extent and trend of impacts from the five key anthropogenic threats on different ecosystems. Almost all the threats are shown to be either continuing in magnitude or increasing for almost all ecosystems. Of the five threats, all, except climate change, already show high to very high impacts on biodiversity in at least some ecosystems. Climate change has had relatively low impact so far but is anticipated to have "very rapidly increasing impacts". Each of the five threats is discussed in more detail below.

Climate Change.

Climate change affects biodiversity in several different ways. Changes in seasonal temperatures or durations can lead to early flowering or egg laying and longer or shorter growing seasons (Secretariat of the CBD 2014).

As global average temperatures increase, climate zones move. This is a gradual process, and in general, species will move with the shifting climate zones. The problem arises when the species cannot relocate. This is particularly problematic for island and mountain creatures. On islands, species movement is limited by island boundaries. On mountains, the extent of warmer zones from the mountain base tends to move higher and higher, reducing the habitable zone for some creatures until the point that there is no habitat left. In other instances, it can be human infrastructure that prevents species from relocating to more suitable climate zones.

When species do manage to relocate, this can be problematic in itself. Ecosystems operate in a natural balance that can be disturbed by the introduction of new species—sometimes to a point of destruction of the ecosystem functioning or the extinction of one or more species. In this way, climate change can be a vector for the introduction of invasive alien species. Changing climates can also provide more favourable conditions for invasive or weedy species. The Intergovernmental Panel on Climate Change predicts that global warming will lead to increased species extinctions, although there is low agreement as to the extent of this (IPCC 2014).

There are two Planetary Boundaries for climate change—one for maximum levels of carbon dioxide in the atmosphere and one maximum change to radiative forcing (the energy balance at Earth's surface). The PB for carbon dioxide is addressed through the PQ for carbon dioxide (see Chap. 8). The PB for radiative forcing is addressed through the PQs for carbon dioxide, methane and nitrous oxide (see Chap. 9), forestland (see Chap. 10), ozone (see Chap. 11), and aerosols (see Chap. 12). As such, the PQ for biodiversity loss does not need to further address this threat to biodiversity.

Pollution.

Pollution is a very broad term that can be simplified by categorizing this as water, land, and air pollution. Water impacts can be further divided into eutrophication (from the release of phosphorus and nitrogen into waterways) and chemical pollution.

The release of nutrients such as nitrogen and phosphorus into the environment poses a very significant threat to biodiversity and ecosystem services globally. Nitrogen and phosphorous fertilizers lower plant diversity and can lead to excessive levels of nutrients in water bodies. The nutrients can lead to dense plant and algal growth (algal blooms) known as eutrophication. Eutrophication can deplete oxygen and solar access to water bodies, wiping out entire ecosystems (Secretariat of the CBD 2014; MEA 2005).

Other pollutants of continuing or growing concern can be generally categorized using the same terminology used in the Planetary Boundary framework—as the release of novel entities into the environment. This includes plastics, in particular their impacts on marine ecosystems, heavy metals, and man-made chemicals which include endocrine disrupters and pesticides, which have been implicated by some studies in damage to pollinating insect and bird populations. Overall, damage from marine oil spills has declined, due to better tanker design and improved navigation, but pollution from pipelines, mainly land-based, has increased due to ageing infrastructure (Secretariat of the CBD 2014).

Air pollution is caused by the release of chemicals, aerosols, and smog precursor gases into the environment. Particulate matter in the atmosphere with particle sizes less than 2.5 μm ($PM_{2.5}$) is considered the most harmful to human health. It is assumed that this can be taken as a proxy measure for human and biodiversity health.

Air pollution is addressed through the PQ for aerosols (see Chap. 12). Water pollution is addressed through PQs for water (see Chap. 13), nitrogen (see Chap. 14), and phosphorus (see Chap. 15). As such, pollution does not need to be further addressed in the PQ for biodiversity. As discussed in Chap. 13, it is very difficult to address all novel entities through a quota approach, and thus, grey water contamination is used as a proxy for novel entities. Further work is needed to consider ways to address novel entities such as plastic more robustly.

The Overexploitation of Species.

The overexploitation of species refers to the harvesting of species at rates higher than population recovery rates of that species. Hunting and especially fishing are considered the two primary drivers. (Secretariat of the CBD 2001). Bushmeat hunting can result in empty forest syndrome which is serious for the forest—75% of tropical trees depend on animals to disperse their seeds. (Secretariat of the CBD 2010)

Overfishing and destructive fishing methods have affected approximately 55% of reefs (Secretariat of the CBD 2014). According to a UN report (2010), more than 80% of global fish stocks are fully exploited or overexploited. The Food and Agriculture Organization (FAO) estimates that we have already reached the maximum wild capture potential for fisheries globally (FAO 2010).

Worm et al. (2009) showed that 63% of 166 assessed fish stocks (the majority of which were well-managed, developed country fisheries) have lower biomass levels than required to obtain maximum sustainable yield (MSY). Costello et al. (2012) found that 64% of fisheries had lower stock biomass than required to support MSY, including 18% that were collapsed.

Large marine protected areas (MPAs) already in place or pending establishment offer opportunities for better protection of coral reefs. Where there are well-enforced MPAs that are coupled with land-based protection, there has been some success in reinstating fish stocks and coral recovery. However only 15% of MPAs have successfully reduced threats from fishing (Secretariat of the CBD 2014).

Birds are often hunted for sport with millions of birds traded internationally each year. Mammals have been hunted for a long time; however, today it is predominantly illegal hunting that threatens mammals—particularly large species.

There are no Pressure indicators which encompass the overexploitation of species. There are various indicators around the appropriation of specific groups of species, in particular fish. For example, the "maximum sustainable yield" for fisheries is aimed at preventing the overharvesting of fisheries. There are no indicators which encompass overexploitation over the broad scale of human impacts.

Introduction of Alien or Feral Species.

Alien or feral species introduced into new environments, whether deliberately or accidentally, have contributed to more than half of the animal extinctions for which the cause is known (Secretariat of the CBD 2014). Globalization has led to a substantial increase in invasive alien species (including disease organisms) (MEA 2005). Species invasions also carry enormous economic costs (Secretariat of the CBD 2014).

The introduction of invasive species has many different pathways. A study by the Convention on Biodiversity CBD (2014) summarized the primary drivers for over 500 invasive species and found over 40 drivers ranging from purposeful release for measures such as erosion control, and hunting, to escape of pets, contamination of international trade objects, and stowaways on container ships. These were categorized into six major groups—release, escape, transport-contamination, transport-stowaway, corridor, and unaided.

Target 9 of the Aichi targets is that by 2020, invasive alien species and pathways are identified and prioritized, priority species are controlled or eradicated, and measures are in place to manage pathways to prevent their introduction and establishment. Eradication programs on islands have been extremely successful, but mainland eradication programs are mostly unsuccessful (Secretariat of the CBD 2014).

Although 55% of countries party to the CBD have policies regarding invasive species, most of these are regarding border control and eradication with very few looking at identifying, prioritizing, and managing pathways of introduction (Secretariat of the CBD 2014). The relationship between trade and biodiversity loss is not simple. More trade leads to increased pressures (through many things but particularly alien species), but also it might allow for more efficient things, therefore reducing net impacts per unit of product (Secretariat of the CBD 2006).

There are no Pressure indicators which encompass the diverse pathways of invasive species introductions.

Land-Use Change.

Land-use change is considered by many to be the greatest threat to biodiversity (Secretariat of the CBD 2010; MEA 2005; Fahrig 2001; Groombridge 1992; Bibby 1994; Ehrlich 1994; Thomas et al. 1994). Kerr and Currie (1995) did a study that looked at different measures of anthropogenic influence on biodiversity loss. Unlike other studies, they did not find that habitat loss was a prime contributor. However, in a study of threats to imperilled species in the United States it was found that habitat destruction and degradation was the greatest threat for 85% of the species analysed.

The biggest driver of land-use change is agriculture (MEA 2005). Population growth is also leading to expansion into formerly natural areas (Fahrig 2001). Hydroelectric power stations are considered to be a sustainable alternative to fossil fuel energy, but they are also a major contributor to habitat loss as hydroelectric dam floods habitats (Secretariat of the CBD 2014). More than one quarter of Earth's terrestrial surface area is already cultivated. A further 10–20% of grassland and forestland is expected to be converted to cultivated land before 2050 (MEA 2005) .

The impacts of land-use change are both direct and indirect. Direct impacts include the destruction, alteration, and fragmentation of habitats. Felling of forests can eliminate habitats for some species. Trawling of sea beds can damage important marine habitats. Indirect impacts include the release of nutrients into waterways and withdrawals of water for irrigation and the segregation of habitats (MEA 2005).

There has been substantial growth in protected areas and many believe such areas are important (Lovejoy 2006). The fact remains that biodiversity is declining even in the face of increasing protected areas (Butchart et al. 2010).

As shown in Sect. 16.2.2, there are a wide range of human-driven factors that influence biodiversity loss that would be very difficult to combine into a single Pressure indicator.[1] However, the high level of interconnectivity between biodiversity loss and other Planetary Boundaries is such that many of the threats are already considered under other Planetary Quotas.

Habitat loss and destruction is considered to some extent through the Planetary Quota for forestland. However, forest is not the only land type that is important. The threat of habitat loss and destruction requires further consideration through a PQ for biodiversity.

16.3.1   A Land-Based Proxy Indicator

The magnitude and diversity of human drivers and pressures with respect to biodiversity loss makes it very difficult to determine 1 or even 2–3 indicators which can address the drivers and pressures holistically. Land use or ecological footprint, water, nitrogen, phosphorous, and carbon impacts have all been used as proxy indicators for biodiversity loss. There have been a few attempts at developing a "biodiversity footprint"; however, these are typically given in State-based indicators and are therefore unsuitable for the Biodiversity Quota (e.g., Hanafiah et al. 2012; Houdet and Germaneau 2014; Moran et al. 2016). There have been some attempts at defining consumption-based biodiversity metrics (e.g., Kitzes et al. 2017). However, these have not yet been developed to the point where they could be used for the Biodiversity Quota.

As previously stated, land use is considered by many to be the greatest threat to biodiversity. It is also an accessible metric. For this reason, the use of land-based indicators as a proxy for biodiversity is common practice. Sustainable Development Goal (SDG) 15 is to "sustainably manage forests, combat desertification, halt and reverse land degradation, and halt biodiversity loss". It is the most explicit SDG with respect to biodiversity loss and includes several land-based indicators in their proposal of suitable indicators to measure this goal including (UN 2015):

* Forest area as a percentage of total land area.
* Forest cover under sustainable forest management.

---

[1] Environmental indicators can be classified as Drivers, Pressures, States, and Impacts using the EU DPSIR framework. Pressure indicators are the type of indicators used for the Planetary Quotas as they can be easily related to human activity and applied at different scales. See Chap. 6 for more detail.

- Percentage of land that is degraded over total land area.

The Ecological Footprint is often used as a proxy indicator for biodiversity health on the basis that it is a measure of how much biologically productive land is used by humans. Some level of overexploitation of marine and terrestrial species is taken into account in this metric (Galli et al. 2014). The problem with using this indicator is that there is little consensus as to an appropriate limit. As discussed in Chap. 5, the term "biodiversity buffer" refers to the amount of global biological capacity that should be left aside for the maintenance of biosphere integrity. The suggested biodiversity buffer ranges greatly.

The Ecological Footprint authors proposed a minimum buffer of 12%, a level that was seconded in the Brundtland Report (Wackernagel et al. 2002; The Brundtland Commission 1987). Soulé and Sanjayan (1998) interviewed 25 conservation leaders, biologists, and agency personnel about what levels would be sufficient. Biologists interviewed suggested that safeguarding 10% could make at least 50% of terrestrial species at risk of anthropogenic extinction (Soulé and Sanjayan 1998). The authors who completed these interviews suggest that 50% is consistent with ecosystem surveys. However, their conclusion is based on very limited studies and low minimum thresholds. In order to estimate the minimum threshold, several studies have been undertaken to estimate how much of existing ecosystems would need to be maintained (by area) in order to maintain all existing species. For example, a study of Australian river valleys showed that 45% of wetland areas were required to represent each species once. However to have each species at least once and represent all wetland types required 75% of wetland areas (Margules et al. 1988). In the Oregon coast range, the authors found that 49% of the ecosystems were required to capture regions of high biodiversity, represent all ecosystems, maintain target species, and provide for connectivity. In Norway 75% of habitat was found to be necessary to protect all plant species in deciduous forests and in Florida 33% to preserve habitats essential for rare and declining species. Fahrig's study "how much habitat is enough" (Fahrig 2001) found that extinction thresholds ranged from less than 1% habitat to over 99% habitat, demonstrating that a single figure for habitat protection is unrealistic.

Galli et al. (2014) have written a paper, *Ecological Footprint: Implications for Biodiversity*, which is intended to demonstrate how Ecological Footprint could be used as an indicator in reducing biodiversity loss; they do not propose a minimum biodiversity buffer. Recommendations for a suitable buffer range widely.

There is no robust way to draw a parallel from the Planetary Boundary limit for extinction rate and a biodiversity buffer using the Ecological Footprint.

In a UNEP report on life-cycle indicators, the need for a scalable indicator to assess the land-use-related impacts on biodiversity was identified and a new indicator proposed (UNEP 2016). The indicator proposed is called the *percentage disappeared fraction* (PDF) of species. This indicator is very similar to the Planetary

Boundary for biosphere integrity—*extinction rate*—as both are expressed in terms of the percentage of extinct (or disappeared) species. The difference between the two is in the calculation. Extinction rate is determined through observation—it is an Impact indicator. In contrast PDF is an estimation based on land-use data—thus it is a Pressure indicator.

This sort of indicator does not address overexploitation or alien invasive species. As discussed previously, it is very hard to assess either of these threats using Pressure indicators. As such, the *percentage disappeared fraction (of species)* is proposed as a proxy indicator for biodiversity.

The purpose of the UNEP report was to propose indicators that allow better consistency in the development and communication of green products. This differs to the purpose of the Quotas in that the Quotas are intended to be the basis of a global Planetary Accounting Framework that can be used for any scale of human activity. In the instance of the UNEP report, there is a little need to account for positive land transformation. As such, all of the "correction factors"—numbers used to convert land transformation to percentage disappeared fraction—are positive (i.e. they lead to biodiversity loss). For the purpose of the Planetary Accounting Framework, further work will be required to determine correction factors for positive transformation which results in biodiversity gains.

16.4   The Limit

The indicator *percentage disappeared fraction* differs from the Planetary Boundary indicator *extinction rate* only in the measurement/calculation method. The unit of measure is fundamentally the same. As such the Planetary Quota for biodiversity is $\leq 1 \times 10^{-4}$/y (PDF).

16.5   Discussion

The current PDF can be estimated to be of an order of magnitude between 1×10^{-3}/y and 1×10^{-2}/y, i.e. 10–100 times greater than the PQ.[2] This indicates that current land use is not amenable to biosphere health. Table 16.2 shows examples of different activities that occur or could occur at different scales and across different sectors to reduce the PDF.

It should be noted that the proposed indicator, percentage disappeared fraction, is a relatively new indicator. It has been developed by a reputable source, i.e. the UNEP. However, work will be needed to assess the reliability of this.

---

[2] These figures are based on global extinction rates as no global PDF has yet been determined.

Table 16.2 Examples of different scales of activity which have or could contribute to achieving the PQ for biosphere integrity

Achieving the planetary quota for biosphere integrity			
	Community	Government	Business
Large	Set up a global organization dedicated to the protection of biodiversity, e.g. *the Global Biodiversity Outlook*	Develop an international treaty for biodiversity management, e.g. *the Convention on Biological Diversity*	Set a commercial guideline for ensuring no business operations have an impact on biodiversity health
Medium	Lobby to prevent land changes with high impacts on biodiversity,e.g. *lobbyists were instrumental in stopping a controversial highway development in Western Australia, Roe 8, which would have led to the destruction of critical local habitats including the Beeliar wetlands*	Design national strategies for habitat zones and biodiversity corridors to connect different habitat zones through urban areas	Review biodiversity practices up and down supply chains
Small	Plant native species	Protect areas of biological importance	
Individual	Plant native species		

16.6   Conclusion

The Planetary Boundary for biosphere integrity has been exceeded by between 10 and 100 times. Biosphere integrity is the most highly connected Planetary Boundary, and almost every Boundary has an impact on biosphere health. The drivers of biodiversity loss are complex, and there is yet to be a metric proposed that deals with these holistically. However, there is a new proxy indicator based on land use that has been developed specifically to estimate human impacts on biodiversity. This is the basis of the Planetary Quota for biosphere integrity which is no more than 1×10^{-4}/y percentage disappeared fraction (of species).

References

ADB (2012) Special chapter: green urbanization in Asia. *Key Indicators for Asia and the Pacific 2012*, 43rd edn. Asian Development Bank, Phillipines

Asafu-Adjaye J (2003) Biodiversity loss and economic growth: a cross-country analysis. Contemp Econ Policy 21:173–185

Bibby CJ (1994) Recent past and future extinctions in birds. Philos Trans R Soc Lond Ser B Biol Sci 344:35–40

Butchart SHM, Walpole M, Collen B, Van Strien A, Scharlemann JPW, Almond REA, Baillie JEM, Bomhard B, Brown C, Bruno J, Carpenter KE, Carr GM, Chanson J, Chenery AM, Csirke J, Davidson NC, Dentener F, Foster M, Galli A, Galloway JN, Genovesi P, Gregory RD, Hockings M, Kapos V, Lamarque J-F, Leverington F, Loh J, Mcgeoch MA, Mcrae L, Minasyan A, Hernández Morcillo M, Oldfield TEE, Pauly D, Quader S, Revenga C, Sauer JR, Skolnik

B, Spear D, Stanwell-Smith D, Stuart SN, Symes A, Tierney M, Tyrrell TD, Vié J-C, Watson R (2010) Global biodiversity: indicators of recent declines. Science 328:1164

CBD (2006) 2nd global biodiversity outlook report. https://www.cbd.int/gbo2/

CBD (2014) Pathways of introduction of invasive species, their prioritization and management. UNEP, Convention on Biological Diversity, Montreal

CBD (2018) History of the convention [Online]. Available: https://www.cbd.int/history/. Accessed 25/01/2018

Chapin IFS, Zavaleta ES, Eviner VT, Naylor RL, Vitousek PM, Reynolds HL, Hooper DU, Lavorel S, Sala OE, Hobbie SE, Mack MC, Diaz S (2000) Consequences of changing biodiversity. Nature 405:234–242

Costello MJ, Wilson S, Houlding B (2012) Predicting total global species richness using rates of species description and estimates of taxonomic effort. Syst Biol 61(5):871–883. https://doi.org/10.1093/sysbio/syr080

Cucek L, Klemes JJ, Kravanja Z (2012) A review of footprint analysis tools for monitoring impacts on sustainability. J Clean Prod 34:9–20

Dietz S, Adger WN (2003) Economic growth, biodiversity loss and conservation effort. J Environ Manag 68:23–35

Ehrlich PR (1994) Energy use and biodiversity loss. Philos Trans R Soc Lond Ser B Biol Sci 344:99–104

Fahrig L (2001) How much habitat is enough? Biol Conserv 100:65–74

FAO (2010) Forestry. Food and Agriculture Organisation of the UN, Rome, Italy

FAO (2018) The state of world fisheries and aquaculture 2018—Meeting the sustainable development goals. Rome

Galli A, Wackernagel M, Iha K, Lazarus E (2014) Ecological footprint: implications for biodiversity. Biol Conserv 173:121–132

Groombridge B (1992) *Global biodiversity : status of the earth's living resources : a report/compiled by the World conservation monitoring centre,* London. Chapman & Hall, London

Hanafiah MM, Hendriks AJ, Huijbregts MAJ (2012) Comparing the ecological footprint with the biodiversity footprint of products.(report). J Clean Prod 37:107

Hansen MC, Potapov PV, Moore R, Hancher M, Turubanova SA, Tyukavina A, Thau D, Stehman SV, Goetz SJ, Loveland TR, Kommareddy A, Egorov A, Chini L, Justice CO, Townshend JR (2013) High-resolution global maps of 21st-century forest cover change. Science 342:850–853

Harper DAT, Hammarlund EU, Rasmussen CM (2014) End Ordovician extinctions: a coincidence of causes. Gondwana Res 25:1294–1307

Houdet J, Germaneau C (2014) Accounting for biodiversity and ecosystem services from an ema perspective: towards a standardised biodiversity footprint methodology. Account Biodiver. https://doi.org/10.4324/9780203097472

IPCC (2014) Climate change 2014: impacts, adaptation, and vulnerability. Part a: global and sectoral aspects. Contribution of working group II to the Fifth assessment report of the intergovernmental panel on climate change [Field CB, Barros VR, Dokken DJ, Mach KJ, Mastrandrea MD, Bilir TE, Chatterjee M, Ebi KL, Estrada YO, Genova RC, Girma B, Kissel ES, Levy AN, MacCracken S, Mastrandrea PR, White LL (eds.)], Cambridge, United Kingdom and New York, NY, USA, Cambridge University Press

Kerr JT, Currie DJ (1995) Effects of human activity on global extinction risk - efectos de la actividad humana sobre el riesgo de extinción global. Conserv Biol 9:1528–1538

Kitzes J, Berlow E, Conlisk E, Erb K, Iha K, Martinez N, Newman EA, Plutzar C, Smith AB, Harte J (2017) Consumption-based conservation targeting: linking biodiversity loss to upstream demand through a global wildlife footprint. Conserv Lett 10:531–538

Lenzen M, Moran D, Kanemoto K, Foran B, Lobefaro L, Geschke A (2012a) International trade drives biodiversity threats in developing nations. Nature 486:109

Lenzen M, Moran D, Kanemoto K, Foran B, Lobefaro L, Geschke A (2012b) International trade drives biodiversity threats in developing nations–supplementary information. Nature 486:109

Lovejoy TE (2006) Protected areas: a prism for a changing world. Trends Ecol Evol 21:329–333

Mace GM, Reyers B, Alkemade R, Biggs R, Chapin FS, Cornell SE, Díaz S, Jennings S, Leadley P, Mumby PJ, Purvis A, Scholes RJ, Seddon AWR, Solan M, Steffen W, Woodward G (2014) Approaches to defining a planetary boundary for biodiversity. Glob Environ Chang 28:289–297

Margules CR, Nicholls AO, Pressey RL (1988) Selecting networks of reserves to maximise biological diversity. Biol Conserv 43:63–76

MEA (2005) Ecosystems and human Well-being: biodiversity synthesis. World Resources Institute, Washinton, DC

Moran D, Petersone M, Verones F (2016) On the suitability of input-output analysis for calculating product-specific biodiversity footprints. Ecol Indic 60:192–201

Naidoo R, Adamowicz WL (2001) Effects of economic prosperity on numbers of threatened species–efectos de la prosperidad económica sobre los números de especies amenazadas. Conserv Biol 15:1021–1029

Rockström J, Steffen W, Noone K, Persson A, Chapin FS, Lambin E, Lenton TM, Scheffer M, Folke C, Schellnhuber HJ, Nykvist B, de Wit CA, Hughes T, van der Leeuw S, Rodhe H, Sorlin S, Snyder PK, Costanza R, Svedin U, Falkenmark M, Karlberg L, Corell RW, Fabry VJ, Hansen J, Walker B, Liverman D, Richardson K, Crutzen P, Foley J (2009) Planetary boundaries: exploring the safe operating space for humanity. Ecol Soc 14:32

Secretariat of the CBD (2001) In: DIVERSITY SOT, B CO (eds) Global biodiversity outlook 1. UNEP, CBD, Montreal

Secretariat of the CBD (2006) Global biodiversity outlook 2. Convention on Biological Diversity, Montreal

Secretariat of the CBD (2010) Global biodiversity outlook 3. Convention on Biological Diversity, Montreal

Secretariat of the CBD (2014) Global biodiversity outlook 4. Convention on Biological Diversity, Montreal

Soulé ME, Sanjayan MA (1998) ECOLOGY: conservation targets: do they help? Science 279:2060

Steffen W, Richardson K, Rockström J, Cornell SE, Fetzer I, Bennett EM, Biggs R, Carpenter SR, de Vries W, de Wit CA, Folke C, Gerten D, Heinke J, Mace GM, Persson LM, Ramanathan V, Reyers B, Sörlin S (2015) Planetary boundaries: guiding human development on a changing planet. Science 347:1259855

The Brundtland Commission (1987) Our common future. World Commission on Environment and Development

Thomas JA, Morris MG, Hambler C (1994) Patterns, mechanisms and rates of extinction among invertebrates in the United Kingdom [and discussion]. Philos Trans R Soc B Biol Sci 344:47–54

Tittensor DP, Walpole M, Hill SL, Boyce DG, Britten GL, Burgess ND, Butchart SH, Leadley PW, Regan EC, Alkemade R, Baumung R, Bellard C, Bouwman L, Bowles-Newark NJ, Chenery AM, Cheung WW, Christensen V, Cooper HD, Crowther AR, Dixon MJ, Galli A, Gaveau V, Gregory RD, Gutierrez NL, Hirsch TL, Hoft R, Januchowski-Hartley SR, Karmann M, Krug CB, Leverington FJ, Loh J, Lojenga RK, Malsch K, Marques A, Morgan DH, Mumby PJ, Newbold T, Noonan-Mooney K, Pagad SN, Parks BC, Pereira HM, Robertson T, Rondinini C, Santini L, Scharlemann JP, Schindler S, Sumaila UR, Teh LS, van Kolck J, Visconti P, YE Y (2014) A mid-term analysis of progress toward international biodiversity targets. Science 346:241–244

UN (2010) Resumed review conference on the agreement relating to the conservation and management of straddling fish stocks and highly migratory fish stocks. United Nations Department of Public Information

UN (2015) United nations sustainable development goals [Online]. Available: https://sustainabledevelopment.un.org/sdgs. Accessed

UNEP (2016) In: Frischknecht R, Jolliet O (eds) Global guidance for life cycle impact assessment indicators. UNEP

Vačkář D (2012) Ecological footprint, environmental performance and biodiversity: a cross-national comparison. Ecol Indic 16:40–46

Wackernagel M, Schulz NB, Deumling D, Linares AC, Jenkins M, Kapos V, Monfreda C, Loh J, Myers N, Norgaard R, Randers J (2002) Tracking the ecological overshoot of the human economy. Proc Natl Acad Sci U S A 99:9266–9271

Watts J. (2019) Human society under urgent threat from loss of Earth's natural life. *The Guradian*

Wilcove DS, Rothstein D, Dubow J, Phillips A, Losos E (1998) Quantifying threats to imperiled species in the United States. Bioscience 48:607–615

Worm B, Hilborn R, Baum JK, Branch TA, Collie JS, Costello C, Fogarty MJ, Fulton EA, Hutchings JA, Jennings S, Jensen OP, Lotze HK, Mace PM, Mcclanahan TR, Minto C, Palumbi SR, Parma AM, Ricard D, Rosenberg AA, Watson R, Zeller D (2009) Rebuilding global fisheries. Science 325:578

Chapter 17
The Imperishable Waste Quota

There is no away
Julia Butterfly Hill

Abstract This PB was first called "chemical pollution" but has now been redefined as "novel entities". However, no limit or control variable has been proposed. This is due to the complexity of measuring and monitoring novel entities. There are too many different chemicals and other entities to sensibly aggregate them into a single measure.

We are proposing two proxy indicators to address novel entities through the Planetary Quotas. The first is the grey water footprint component of the water Quota. We do not propose a specific limit for grey water. Rather, we acknowledge that water pollution will be limited through the grey water component of the water Quota.

The second proxy indicator we are proposing is *net imperishable waste* which is about long-term pollutants such as plastic and heavy metals. There is evidence that we have already gone beyond the global limit for imperishable waste assimilation. As such, we propose that this Planetary Quota is *net imperishable waste ≤0 kgs*.

17.1 Introduction

There are hundreds of thousands of man-made chemicals, materials, and substances. The potential effects of these substances are often poorly understood. Chlorofluorocarbons (CFCs) are an example of man-made chemicals that were thought to be a breakthrough for many human needs—in particular refrigeration—as they were seemingly so much safer than previously used refrigerants. These substances that were touted for being harmless turned out to have serious, unexpected, global effects—thinning the ozone layer to the point that every spring there is a large area with almost no ozone at all (see Chap. 11). The damaging end products from these man-made chemicals such as mercury, originally included as "chemical pollutants" by the Planetary Boundaries scientists are now referred to as "novel entities" to encompass all man-made entities.

© Springer Nature Singapore Pte Ltd. 2020
K. Meyer, P. Newman, *Planetary Accounting*,
https://doi.org/10.1007/978-981-15-1443-2_17

Although it is included in the Planetary Boundaries framework, novel is the one Planetary Boundary for which no limit, or even control variable has been proposed.

This chapter begins by discussing the reasons to include novel entities as a Planetary Boundary despite the challenges in determining a control variable or limit for these. It goes on to discuss why we have included imperishable waste as a proxy indicator and to make the case for a zero limit. The chapter concludes with a discussion on the status quo and the actions that may be required in order to meet this Planetary Quota.

17.2   Background

Natural resource scarcity is gaining growing policy attention in light of emerging scarcity of key resources (Commission 2011; Council 2008). The "material footprint" (MF), i.e. the amount of raw material associated with human activity each year, ranges from 3.7 tonnes per capita (t/cap) in India to 35 t/cap in Australia (Wiedmann et al. 2015). As a global population, we currently produce more than 2 billion tons of waste every year (Kaza et al. 2018). The World Bank have projected that without urgent action this is likely to increase to 3.4 billion tonnes by 2050 (Kaza et al. 2018). In their report, "What a Waste", their recycling rates remain low, ranging from 16 to 50% (Kaza et al. 2018).

Of all the plastic ever produced, approximately half has been produced since the year 2000, and over 75% is already waste (Dalberg 2019). It is estimated that 25% of plastic produced is now "in nature" as either land, freshwater, or marine pollution (de Souza Machado et al. 2018). Tiny particles of plastic "macro, micro, and nano" plastics are becoming ubiquitous contaminating our soils, freshwater, and oceans (Murphy et al. 2016) and therefore being ingested by each and every one of us: a couple of particles as you enjoy a pint of beer after work, another as you add a pinch of salt to your fries; prefer a healthier lifestyle? That's ok you are not missing out—tap water has up to 60 particles of plastic per litre (Kosuth et al. 2018). The health impacts of ingesting microplastics are still largely unknown (Kosuth et al. 2018). However, we do know that there are serious impacts on wildlife, who are ingesting not only microplastics but also much larger plastics found in nature. A popular example touted through social media channels is the likeness of a floating plastic bag to a jellyfish as depicted in Fig. 17.1. The creation of plastics is very carbon intensive—contributing to approximately 4% of global oil and gas demands each year (IEA 2018). The natural capital cost of plastic has been estimated at US$8 billion (Kaza et al. 2018).

It is estimated that there are between 80,000 and 100,000 chemicals on the market (Egeghy et al. 2012, Commission of the European Communities 2001, U.S. Environmental Protection Agency 2008). There is very limited understanding of the potential impacts on the health and wellbeing of the planet of many of these compounds (Daly 2006). However, we do know that many of them have far-reaching impacts on our planet (Rockström et al. 2009b). Biodiversity can be affected directly, where excessive pollutants cause species loss, and indirectly, by increasing species

Fig. 17.1 Jellyfish or shopping bag (Seegraswiese 2013) (CC3.0). Creative commons attribution-share alike 3.0 unported

vulnerability to other stresses such as climate change. Most industrial chemicals are produced from petroleum, thereby producing carbon dioxide and thus contributing to climate change.

17.3   A Planetary Boundary with No Boundary

The ninth Planetary Boundary in the first Planetary Boundary publication was for chemical pollution (Rockström et al. 2009a). In this publication, chemical pollution was referred to as pollution from radioactive compounds, heavy metals, and other organic compounds developed by humankind.

There was neither an indicator nor a proposed limit (Rockström et al. 2009b) although the authors included the following description as an example control variable "emissions, concentrations or effects on ecosystem and Earth-system functioning of persistent organic pollutants (POPs) plastics, endocrine disruptors, heavy metals, and nuclear waste".

Chemical pollution was included in the Planetary Boundaries framework because of the adverse effects on both human and ecosystem health. These effects have predominantly been seen at local and regional scales. The authors argue that a global boundary exists based on two rationales:

1. A direct global impact on the physiological development of humans and other organisms which changes ecosystem function or structure.

2. Indirect impacts on other Boundaries—for example, weakening species resilience to withstanding the impacts of climate change.

In the 2015 update of the PBs, they redefined this Planetary Boundary from chemical pollution to "novel entities" (Steffen et al. 2015). They defined these as "new substances, new forms of existing substances and modified life-forms that have the potential for unwanted geophysical and/or biological effects." They identified three key risks of novel entities at a global level: persistence, mobility, and impacts on vital Earth-system processes or sub-systems.

The problem with attempting to set a Planetary Boundary for novel entities is that there are over 100,000 man-made chemicals currently in circulation (Egeghy et al. 2012, Commission of the European Communities 2001, U.S. Environmental Protection Agency 2008) not including nonmaterial and plastic polymers that degrade into microplastics. There is no existing mechanism with which to amalgamate these into a single control variable. It would not be possible to set a limit for every chemical. Rockström et al. (2009b) have proposed a twofold approach. Firstly to focus on persistent pollutants that can travel long distances through the ocean or atmosphere such as mercury. The other is to identify unacceptable long-term and widespread impacts. The first part of this approach lends itself to one or several Pressure indicators. The latter would be more likely to be a State or Impact indicator such as reduced or failed reproduction, neurobehavioral deficits, and compromised immune systems (Rockström et al. 2009b).

17.4   An Indicator for Novel Entities

There are efforts underway enabling global impacts of chemicals to be considered to assess whether or not chemicals pose a global threat (Macleod et al. 2014). This is a binary system where chemicals reviewed are classified either as posing a Planetary Boundary threat or not. The criteria used for assessment are:

- The chemical has an unknown disruptive effect on a vital Earth-system process.
- The chemical has a disruptive effect that is not discovered until it is a problem at the global scale.
- The effect is not readily reversible.

Any chemical fulfilling all three criteria is classified as a threat. This work was not considered suitable as the basis for a Planetary Boundary (Steffen et al. 2015); however, they acknowledged that the approach could be useful in due course.

The reason it would not be suitable as a Quota variable is that the threat could not be confirmed until it is too late as the criteria include "unknown" and "undiscovered" disruptive effects.

However, the lack of a specific Planetary Boundary does not mean that novel entities should not be considered within the Planetary Quotas. It is common practice to use water pollution as a proxy measure for chemical pollution (Bjørn et al. 2014).

This is typically done by determining the dilution factor—the amount of water that would be needed to assimilate any pollution.

This proxy indicator is already imbedded within the Planetary Quota for water (see Chap. 13). However, this indicator does not allow for all novel entities to be considered. Entities which do not make it to water bodies, and those which cannot be diluted (for example microplastics), are not accounted for in a water pollution metric. As such, we are also proposing a proxy indicator for solid waste: net imperishable waste, measured in kilograms or tonnes of waste.

There are several limitations of this approach. To begin with it does not capture all novel entities. Moreover, the weight of novel entities is not the best indicator of their risk—for example, a small amount of mercury can do a lot more harm than a large amount of non-toxic waste. However, we argue that it is the best currently available proxy indicator that can be used to begin to manage novel entities.

17.5 A Limit

Given that there is Planetary Boundary limit, or limit in the literature for maximum chemical water pollution, we do not propose a specific limit for the grey water footprint. Rather, it is our assumption that incorporating this limit into the global water limit leaves humanity to determine how best to manage water resources and thus water pollution.

There is no specific limit proposed in the literature for a sustainable level of imperishable waste. However, there is substantial evidence that we are beyond the limit. For example, 83% of tap water samples from 12 nations have been found to be contaminated with plastic (Tyree and Morrison 2017); methane from landfills and wastes contributes approximately 23% of global methane emissions (Ciais et al. 2013); most fish which are high in the food chain now contain high levels of heavy metals such as mercury (U.S FDA 2017). We thus propose that the PQ for novel entities should be net imperishable waste ≈ 0 kg/year.

The choice of net rather than gross waste is to allow environmental impact assessment results to show negative imperishable waste disposal. In this way, activities such as landfill mining which result in a net removal could be encouraged. Value could be assigned to such activities to allow for trading of impacts within a global cap. Further work should be undertaken to determine whether a zero limit is sufficient.

17.6 Discussion

Humans currently produce more than 2 billion tons of waste each year (Kaza et al. 2018). Much of this is likely to be perishable waste. However, the impacts of perishable waste when disposed of in landfill can be considered equivalent to imperishable waste in landfill. Food waste appropriately composted produces useful

Table 17.1 Examples of different scales of activity which have or could contribute to achieving the PQ for imperishable waste

Achieving the Planetary Quota for biosphere integrity			
	Community	**Government**	**Business**
Large	Set up a global organization dedicated to supporting waste free-living	Develop an international agreement regarding waste management	Develop innovative solutions to more efficient landfill mining
Medium	Share lessons learnt in successful waste reduction programmes with similar communities	Develop legislation and policy to significantly reduce waste production	Reduce the use of chemicals and plastics in business operations
Small	Set up community composting centres	Incentivise waste reductions	Switch from single-use plastics to recyclable or compostable alternatives
Individual	Make purchasing decisions that minimize waste production	Take a stand against single-use plastics	Lobby the business to measure and manage imperishable waste

fertilization for soils. In contrast, food waste in landfills generates methane, a dangerous greenhouse gas (see Chap. 9).

Table 17.1 lists examples of the sorts of activity that may be needed to achieve this Quota.

17.7   Conclusion

There is no Planetary Boundary control variable or limit for novel entities. Nonetheless novel entities are included in the Planetary Boundary framework due to their potential impact on the functioning of the Earth system. Two mechanisms exist within the Planetary Quotas to address imperishable waste. The grey water footprint component of the water footprint and the Quota for imperishable waste.

While there is no specific limit in the literature for imperishable waste, there is evidence to suggest that we are beyond the limit. As such, we have proposed a PQ for imperishable waste as net imperishable waste ≤ 0 kgs.

References

Bjørn A, Diamond M, Birkved M, Hauschild MZ (2014) Chemical footprint method for improved communication of freshwater ecotoxicity impacts in the context of ecological limits. Environ Sci Technol 48:13253

Ciais P, Sabine C, Bala G, Bopp L, Brovkin V, Canadell J, Chhabra A, Defries R, Galloway J, Heimann M, Jones C, le Quéré C, Myneni RB, Piao S, Thornton P (2013) Carbon and other biogeochemical cycles. In: Stocker TF, Qin D, Plattner G-K, Tignor M, Allen SK, Boschung J, Nauels A, Xia Y, Bex V, Midgley PM (eds) Climate change 2013: the physical science basis.

Contribution of working group I to the fifth assessment report of the intergovernmental panel on climate change. Cambridge University Press, Cambridge, United Kingdom and New York, NY, USA

Commission, E (2011) Roadmap to a resource efficient europe. COM (2011) 571 Final. EU Commission, Bruxelles

Commission of the European Communities (2001) White paper: strategy for future chemicals policy (COM2001 88 final)

Council NR (2008) Minerals, critical minerals, and the US economy. National Academies Press, Washington

Dalberg (2019) Solving plastic pollution through accountability. Gland, Switzerland

Daly G (2006) Bad chemistry. onearth

de Souza Machado AA, Kloas W, Zarfl C, Hempel S, Rillig MC (2018) Microplastics as an emerging threat to terrestrial ecosystems. Glob Chang Biol 24:1405–1416

Egeghy PP, Judson R, Gangwal S, Mosher S, Smith D, Vail J, Cohen Hubal EA (2012) The exposure data landscape for manufactured chemicals. Sci Total Environ 414:159–166

IEA (2018) Oil 2018–analysis and forecasts to 2023. Market report series. International Energy Agency

Kaza S, Yao LC, Bhada-Tata P & van Woerden F (2018) What a waste 2.0 : a global snapshot of solid waste management to 2050. Washington, DC, World Bank. © World Bank. https://openknowledge.worldbank.org/handle/10986/30317 License: CC BY 3.0 IGO

Kosuth M, Mason SA, Wattenberg EV (2018) Anthropogenic contamination of tap water, beer, and sea salt. PLoS One 13:e0194970

Macleod M, Breitholtz M, Cousins IT, de Wit CA, Persson LM, Rudén C, Mclachlan MS (2014) Identifying chemicals that are planetary boundary threats. Environ Sci Technol 48:11057

Murphy F, Ewins C, Carbonnier F, Quinn B (2016) Wastewater treatment works (WwTW) as a source of microplastics in the aquatic environment. Environ Sci Technol 50:5800–5808

Rockström J, Steffen W, Noone K, Persson A, Chapin FS, Lambin E, Lenton TM, Scheffer M, Folke C, Schellnhuber HJ, Nykvist B, de Wit CA, Hughes T, van der Leeuw S, Rodhe H, Sorlin S, Snyder PK, Costanza R, Svedin U, Falkenmark M, Karlberg L, Corell RW, Fabry VJ, Hansen J, Walker B, Liverman D, Richardson K, Crutzen P, Foley J (2009a) Planetary boundaries: exploring the safe operating space for humanity. Ecol Soc 14:32

Rockström J, Steffen W, Noone K, Persson A, Chapin FS III, Lambin E, Lenton TM, Scheffer M, Folke C, Schellnhuber HJ, Nykvist B, de Wit CA, Hughes T, Van der Leeuw S, Rodhe H, Sörlin S, Snyder PK, Costanza R, Svedin U, Falkenmark M, Karlberg L, Corell RW, Fabry VJ, Hansen J, Walker B, Liverman D, Richardson K, Crutzen P, Foley J (2009b) Planetary boundaries: exploring the safe operating space for humanity. Ecol Soc 14:32

Seegraswiese (2013) Plastic_bag_jellyfish.png. Wikimedia

Steffen W, Richardson K, Rockström J, Cornell SE, Fetzer I, Bennett EM, Biggs R, Carpenter SR, de Vries W, de Wit CA, Folke C, Gerten D, Heinke J, Mace GM, Persson LM, Ramanathan V, Reyers B, Sörlin S (2015) Planetary boundaries: Guiding human development on a changing planet. Science 347:1259855

Tyree C, Morrison D (2017) Invisibles: the plastic inside us. Orbmedia.org

U.S FDA (2017) Mercury levels in commercial fish and shellfish (1990–2012) [Online]. U.S. Department of Health and Human Services. Available: https://www.fda.gov/Food/FoodborneIllnessContaminants/Metals/ucm115644.htm. Accessed 6 June 2018

U.S. Environmental Protection Agency (2008) Chemical hazard data availability study: what do we really know about the safety of high production volume chemicals? Climate and Society: Lessons from the Past 10 000 Years, 476–482

Wiedmann TO, Schandl H, Lenzen M, Moran D, Suh S, West J, Kanemoto K (2015) The material footprint of nations. Proc Natl Acad Sci 112:6271–6276

Part III
A New Paradigm of Environmental Management

Chapter 18
The Planetary Accounting Framework

You can't manage what you don't measure
Peter Drucker

Abstract There is a need for a poly-scalar approach to Earth-system management. Such an approach should be one which is integrative across different scales, sectors, and timeframes. It should not be controlled by a single body, but could be implemented through governance, privatization, or self-organized management, that is coordinated by a general system of rules which have different mechanisms at different centres of activity. Planetary Accounting enables a poly-scalar approach. It extends the concepts of carbon accounting and science based targets accross all of the Planetary Boundaries. It enables the comparison of the environmental footprints of human activity to a share of scientific global limits. It can be used to report progress towards an "end goal" for sustainability. However, its unique value is its utility as a decision making tool at all scales of human activity. Planetary Accounting creates a framework for innovation and transformative change towards human activity within the Planetary Boundaries.

In order for Planetary Accounting to be possible, we derived a new set of global limits, the "Planetary Quotas" that connect the Planetary Boundaries to environmental accounting in a way that can be scaled and used as the basis of decision-making. Planetary Accounting allows people to quantify what needs to be done to return to and maintain planetary health.

The Planetary Accounting Framework shows how to use the Planetary Quotas and environmental accounting methods to calculate environmental parameters for different scales and types of activity and to compare these to the status quo. One potential output of Planetary Accounting is an *impact balance statement* which shows the environmental credits or deficits of an activity in each environmental currency.

Planetary Accounting can be used to inform policy and governance, business operations, legislation, design and technology, and behaviour change programs. It is the basis for a new not-for-profit research centre—the Planetary Accounting Network (PAN). Planetary Accounting has the flexibility and high resolution needed to form the general system of rules for a poly-scalar approach to Earth-system management.

© Springer Nature Singapore Pte Ltd. 2020
K. Meyer, P. Newman, *Planetary Accounting*,
https://doi.org/10.1007/978-981-15-1443-2_18

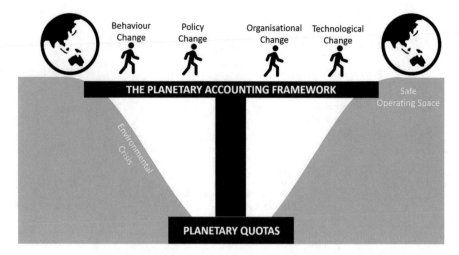

Fig. 18.1 The Planetary Quotas provide the foundation for the Planetary Accounting Framework. The framework is a platform that could enable change at all levels of human activity

18.1   Introduction

In the first section of this book, we identified a gap—the need for a new approach to Earth-system management—a poly-scalar approach. We showed that in order for such an approach to be possible, a new set of global limits would be needed, based on the Planetary Boundaries, but in metrics that make sense at smaller scales and which can be directly compared to human activity.

The second section of this book described how we derived the Planetary Quotas and provided a detailed summary of why each of the Quotas is needed and what actions might be needed to improve our management of them.

This final section of the book explains how Planetary Accounting works. It shows how the PQs form the foundations of Planetary Accounting, a platform for behavioural, policy, technological, and organizational change (see Fig. 18.1). It begins with an overview of the Planetary Quotas and how we are currently performing against these at a global level. It then describes how to use environmental accounting methods and the Planetary Quotas to determine science-based targets for any scale of activity and to compare these with the status quo.

The chapter goes on to discuss allocation methodologies and show how the PAF has the flexibility to be applied across different methodologies rather than necessitating the selection of one over another. This chapter also includes an overview of what we perceive as the key strengths and weaknesses of our approach.

PLANETARY QUOTAS

Fig. 18.2 The Planetary Quotas: Scalable global limits for human-induced environmental impacts in environmental currencies such as carbon emissions, reforestation, and water consumption

The chapter concludes by summarizing the ongoing and planned future work in this space.

18.2   The Planetary Quotas

Figure 18.2 depicts the Planetary Quotas—ten environmental currencies with global limits. Each PQ can be allocated to different scales of activity including individual, city, business, sector, national, and global scales or shares.

Table 18.1 summarizes the Planetary Quotas, showing the indicators and preliminary limits for each. Table 18.2 shows the global status against each of the PQs. Seven of the ten PQs are currently being exceeded, the PQs for carbon, methane and nitrous oxide (MeNO), forestland, biodiversity, nitrogen, and phosphorus and Montreal gas emissions. Current water consumption is at the PQ limit for water. Current impacts have not yet been assessed for the aerosol emissions.

It is important to note that exceeding a Planetary Quota is not the same as exceeding a Planetary Boundary. The PQs do not relate to the state of the environment. Rather, they are about the annual impacts of human activity. As such, it is possible to be exceeding a PQ while remaining within corresponding PBs and vice versa. What we know from the PQs is the direction of change. If, for example, we were able to operate within the PQs for carbon, we would be moving towards the PB for atmospheric carbon dioxide. However, it would take a long time to return to within this PB. Likewise, although we have not currently exceeded the PB for water, if we operate beyond the PQ for water, it is likely that we will exceed the PB in time.

In Chap. 7, we used the example of health to explain the relationship between the Planetary Boundaries (the parameters for a healthy planet) and the Planetary Quotas (the prescription for a healthy planet). To improve the health of the patient, a doctor might suggest a diet—with maximum calorific intake per day. She might suggest a minimum duration of exercise per day or an absolute reduction of cigarette consumption to zero. These ongoing guidelines will not instantly fix the patient's health, but over time, his health will improve, and eventually, he should return to a healthy benchmark.

Table 18.1 The Planetary Quotas

Environmental currency	Planetary quota (control variable and limit)	Description of control variable
Carbon	Net carbon emissions ≤ -2.6 GtCO$_2$/yr	Net CO$_2$ emissions (including land use and land-use change emissions)
Methane and nitrous oxide (MeNO)	MeNO emissions ≤ 5.4 GtCO$_2$e/yr	Net warming potential of methane and nitrous oxide emissions to the atmosphere (in CO$_2$e)
Forestland	Deforestation rate ≤ -11Mha/yr	Net deforestation rate
Aerosols	$0.04 \leq$ AODe ≤ 0.1	Air quality impacts of emissions of aerosols and precursor gases expressed in *equivalent aerosol optical depth*
Ozone	Montreal gas emissions ≈ 0 ODP Kg/yr	Emission of gases controlled or due to be controlled under the Montreal Protocol in terms of ozone-depleting equivalence
Nitrogen	Net nitrogen release ≤ 62 TgN/yr	Net release of reactive nitrogen to the environment
Phosphorous	Net phosphorous release ≤ 11 TgP/yr	Net release of phosphorus to the environment
Water	Net water consumption ≤ 8500 km^3/yr	Net green, blue, and grey water footprint[a]
Biodiversity	PDF $\leq 1 \times 10^{-4}$/yr	Net percentage disappearing fraction of species due to land occupation and transformation
Imperishable waste	Net imperishable waste ≈ 0 kgs	Net waste to landfill

Notes

[a]*Green* = rainwater, *blue* = surface and groundwater, *grey* = the amount of freshwater required to dilute contaminated water to acceptable standards

Like the exercise and diet, the PQs will not immediately return the planet to the healthy state described by the PBs. However, if followed, we know we are going in the right direction. Moreover, we can predict early on that we are heading towards PB limits that have not yet been exceeded if we are continually exceeding PQs. By continuing to operate within the Quotas, humanity can feel confident that we are not pushing the planet beyond its limits.

There is not timeframe assigned to the Planetary Boundaries. There is no pathway of incremental change proposed by the authors. Rather their research defines the end goal. This is intentional. It is up to each person, group, business, or sector to define their own transition pathways to operate within the PBs. Likewise, the purpose of the PQs is to allow humanity the freedom and flexibility to determine the best way to operate within the safe operating space. Where we have had to select a timeframe within which to respect the PBs, notably for the PQs for carbon dioxide and Me-NO, we have chosen the end of this century, i.e. the soonest date considered possible within the academic literature. Of the ten PQs, only the PQ for carbon dioxide has a date at which it must be respected in order for it to remain at the point at which it is currently set. For the remaining PQs, we have not proposed a specific

Table 18.2 Each of the Planetary Quotas is shown against the estimated current global status showing five of the Quotas are currently exceeded, one is on the threshold, and the remaining two are unknown

Planetary quota currency	Limit	Estimate of current global status
Carbon emissions	≤ -7.3 GtCO$_2$/y	36 GtCO$_2$/y[a]
Me-NO emissions	≤ 5.4GtCO$_2$e/y	11 GtCO$_2$e/y[b]
Deforestation rate	≤ -11 Mha/y	6.5 Mha/y[c]
Aerosol emissions	≥ 0.04, and ≤ 0.1	Data not available but likely exceeded[d]
Montreal gas emissions	≈ 0 ODP tonnes/y	32,000 ODP t/y[e]
Nitrogen release	≤ 62 TgN/y	150 Tg/y[f]
Phosphorous release	≤ 11 TgP/y	22 Tg/y (Steffen et al. 2015)
Water consumption	≤ 8500 km^3/y	8500 km^3/y[h]
Biodiversity	≤ 1E-4/y	$1 \times 10^{-2} - 1 \times 10^{-3}$/y[h]
Imperishable waste	≈ 0 kgs/y	2 billion t/y[i]

Notes
[a](World Bank 2019)
[b]Derived from World Bank (2019)
[c](FAO 2016)
[d]In 2016, 92% of the world's population lived in areas that exceed the World Health Organization ambient air quality guidelines (WHO 2016). This suggests this Quota (which is based on these guidelines) has been exceeded
[e](UN 2016)
[f](Steffen et al. 2015)
[g](Hoekstra 2017)
[h]Based on background extinction rate of 100–1000 extinctions per million species per year (Steffen et al. 2015)
[i](Kaza et al. 2018)

time within which they must be respected. Every year that we exceed them, the risk of an irreversible departure from a Holocene-like state of the Earth system increases. To return to the example of health—an overweight person is at risk of irreversible weight-related problems such as heart failure for as long as he is overweight. The sooner he begins to diet, the lower the risk. The inclusion of a timeframe for each PQ should be considered in future work.

It should be noted that unlike for the PBs, no "zone of uncertainty" has been included for the PQs. The zone of uncertainty is included in the Planetary Boundary framework to account for the fact that the science is uncertain. The PQs are intended to show how to operate within the PBs. For this reason, the PQ limits are set according to the lower limits in the Planetary Boundary framework. Future work should include estimations of uncertainty around the PQ limits.

The Earth system is dynamic and the rate of increase in scientific understanding of its processes and limits is high. There is not time to wait until we have a perfect understanding of the system or its limits before we act to try to operate within these; this may never eventuate. The indicators and limits shown in this book are intended to be preliminary. It is our intention that, like the Planetary Boundaries, these are subjected to scrutiny, discussion, and analysis and that they are regularly reviewed and updated over time as we advance in our collective knowledge and understanding.

18.3   The Planetary Accounting Framework

Figure 18.3 summarizes how Planetary Accounting works for different scales and purposes. The left-hand side shows the inputs and the right-hand side shows the outputs. The inputs are both top-down—scaling the Planetary Quotas to the scale of assessment—and bottom-up—using environmental impact assessment methods to estimate impacts in each environmental activity.

A poly-scalar approach was defined in Chap. 4 as one which is *integrative across different scales, sectors, and timeframes, that is not controlled by a single body, but which could be implemented through governance, privatization, or self-organized management, that is coordinated by a general system of rules which have different mechanisms at different centres of activity.*

The diagram shows how Planetary Accounting can be used as this general system of rules. The colours show how the PAF addresses each of the core elements described in Sect. 18.1 of this book:

- Orange: Earth's limits—the Planetary Boundaries (via the Planetary Quotas)—from Chap. 3.
- Green: the poly-scalar Earth-system management approach—from Chap. 4.
- Blue: the environmental accounting system against global limits—from Chap. 5.

In addition to these three elements, the yellow boxes show how the PAF has different mechanisms at different scales.

Each box shown in Fig. 18.3 is discussed in the following sections.

18.3.1   Bottom-Up

The lower left quadrant of Fig. 18.3 shows the inputs required for the bottom-up portion of the accounting procedures. There are three stages in this quadrant: determining the purpose of the accounts, selecting the scope of environmental assessment, and performing the environmental assessment.

18.3.1.1   Purpose

For Planetary Accounting to be used in a poly-scalar approach to managing the Earth system, it must have a high level of flexibility in the way it is applied with different mechanisms at different scales and for different purposes. Different accounting procedures should be used depending on the purpose of the accounts. For example, a city scale might have several different purposes for which the Planetary Accounting Framework could be used such as the assessment of:

- Per capita impacts of the city's inhabitants for comparison with other cities.

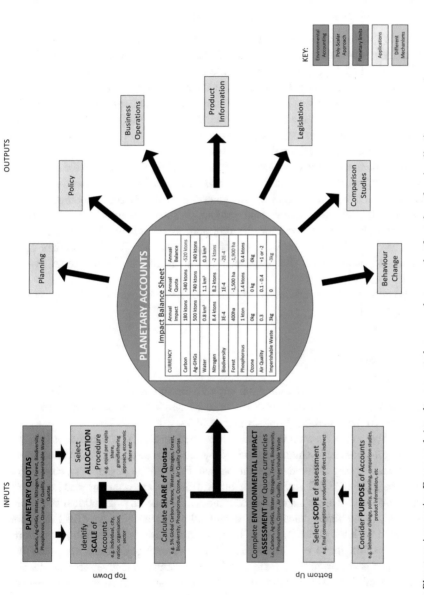

Fig. 18.3 The Planetary Accounting Framework (numbers reported are a random sample of numbers for visualization purposes)

- Impacts within the city government's jurisdiction to control—to inform local policy and/or planning.
- Future impacts under different scenarios—e.g. the likely impacts after a new public transport system was developed or a new minimum building standard was implemented—to support decision-making.

When looking at the per capita impacts of the inhabitants for comparison with other cities, it would be more relevant to consider the consumption and environmental impacts of each person—whether these occur within the city boundaries or not. For example, the impacts of meat consumed by the city residents should be considered whether the meat was produced within the city boundaries or not. On the other hand, if the city was trying to understand the impacts of infrastructure and activity within their jurisdiction, meat produced outside the city limits would not be relevant and should be excluded from the accounts.

The inclusions and exclusions can make a very big difference to the results. For example, in Sweden the emissions produced within the Swedish borders are approximately 50–60% of the emissions associated with the Swedish population's consumption (Dawkins et al. 2019). Both production and consumption accounts provide information that is useful, but for different purposes. For example, it does not make sense to divide the total emissions produced within the Swedish borders by the number of inhabitants if the purpose is to understand the behaviour of Swedish people.

The purpose of the accounts must thus be determined prior to defining the scope of assessment.

18.3.1.2   Scope

Scope is a term used in environmental impact assessments to describe which impacts are included and which are excluded. There are two key considerations when defining scope for environmental impact assessments.

The first is to determine whether the assessment is *consumption-based* or *production-based* assessments as defined by the purpose above. The inclusions and exclusions are very different between the two. For example, *production* calculations would include all GHG emissions resulting from activities taking place within the jurisdiction under assessment, for example, within the city or national borders. In contrast, *consumption* calculations would take into account all GHG emissions associated with the consumption of the people residing in the country or nation—regardless of the location of the emissions themselves. The difference between these two scopes is shown in the example of Sweden's emissions given in Sect. 18.3.1.1 This type of scope selection is most relevant for scales such as city or national accounts.

The second is to determine the boundaries of influence. This type of scope selection is most relevant for businesses and households. In some instances, it may be reasonable to consider only the impacts under the direct control of the party under assessment. In the example of a household or business, this would include electricity and water consumption. In other instances, a broader view is more appropriate, for example, the impacts of the products and/or raw materials purchased or used by the business or household may need to be understood. Having a clearly defined purpose for the assessment helps to identify the scope.

It is important that a clearly defined scope is used for any instances where the assessments are going to be compared with one another. Inclusions and exclusions are not necessarily immediately obvious. Take the example of Planetary Accounts for an individual. Should work-related impacts such as the electricity consumed by an employee's computer be attributed to the employee? To the business owner? To the final consumer of the business output? Should this differ in the instance that the employee works on a farm or an office? To whom should the impacts of food be attributed? The producer or the consumer?

There is not a right or wrong answer to such questions. The answers will depend heavily on the purpose of the accounts—and there can be many sets of accounts for different purposes. For example, if the purpose of a set of accounts is to compare and evaluate behaviour and consumer choices, then it would make sense to attribute impacts to the final consumer. However, if the purpose is to understand the impacts that an individual can influence, then the impacts that person can influence at their workplace should probably be considered.

As the fundamental premise of Planetary Accounting is to compare impacts to a share of global limits, it is also important to consider double counting and missed impacts in the scope definition. If, for example, an individual's impacts were to be compared against an equal per capita share of the Planetary Quotas, then one must consider how to deal with public service impacts such as the impacts of healthcare, national security, and local infrastructure. There are many ways public service impacts could be dealt with. For example, it could be assumed that 25% of an individual's share of the Quotas would need to be used for public services, and the individual's impacts would thus be compared to the remaining 75% of a per capita share. Alternatively, the actual impacts of public services in the individual's region could be determined and a portion of these attributed to the individual.

This is intended as an example only. Planned future work includes the development of formal Planetary Accounting standards to include this level of detail for different types of accounts, in the same way that there are formal standards for life-cycle assessments.

18.3.1.3 Environmental Impact Assessment

The environmental impacts within the defined scope should then be calculated. For example, the emissions of CO_2, the type and amount of land used, the water consumed, and the nitrogen released to the environment would all need to be determined. In the instance that both direct and indirect impacts are included in the scope, the amount of time and cost associated with the data collection for this process is likely to be prohibitive at first. Future work should include the development of basic databases that can be used to access standard impact factors in the absence of more specific data. As more activities are assessed, a wiki-type database could be used to capture this information through crowdsourcing so that the availability of information improves over time (see Box 18.1). Companies that see value in the concept of Planetary Accounting may start to disclose this information to consumers. There could even be an opportunity for a *Planetary Facts* labelling scheme in the same way that nutritional facts are now mandated on most foods in many countries (this is discussed further later in this Chapter).

Further work should include the development of environmental assessment standards for Planetary Accounting.

Box 18.1 Crowdsourced Data

I first discovered *mapmyfitness.com*, a health and exercise app, in 2010. The purpose of the app is to track calories eaten against exercise to aid in weight loss. In 2010, when I used the app, the calories in basic foods, such as eggs, flour, oil, and vegetables, were embedded within the app. There was some data for specific meals, such as lasagne, and for specific brands and products, such as a Snickers bar. However, in the absence of data, the user would upload their own. For a home-made meal, the app requested the total quantities of each food type to be uploaded, and then the user could select what fraction of the meal they had consumed. For products with nutritional facts labels, there was the option to copy these details directly. All of the data entered was then stored and made available to any user. A verification process allowed other users to confirm the accuracy of inputs.

In 2017, when I returned to the app to track my calories, almost every meal or product I entered was already available in the apps database. This had all been entered and verified by users—i.e. it had been crowdsourced.

18.3.2 Top-Down

The upper left quadrant of Fig. 18.3 shows the inputs required for the top-down portion of the accounting procedures. This is the process of scaling the Planetary Quotas from global limits to an appropriate share or allocation.

18.3.2.1 Scale of Accounts

The first information needed is the scale of the accounts. The accounts might be national, for a city or individual or for a business or sector. The scale of accounts must be understood to calculate the appropriate share of the PQs. The purpose of the accounts will help to determine the scale (see Sect. 18.3.1.1).

18.3.2.2 Allocation Procedure

The Planetary Quotas resolve the mathematics of apportioning shares of the operating space to different scales of human activity. However, distributing Earth's finite resources among past, present, and future generations is not simply a question of mathematics. It is a question of ethics, morals, and beliefs. It is a political debate. If our forefathers used more than their "fair share" of resources, should we be penalized for these decisions? How far should environmental practices alter the lives of people today for the benefit of future generations? Should poor countries, with limited access to basic human rights, be allowed to emit fossil fuels to "catch up" when we know that such emissions could push the Earth system into a new paradigm?

We do not attempt to resolve these issues here. However, Planetary Accounting cannot be practised without the step of determining a share of the safe operating space; thus, in any Planetary Accounts, some allocation method must be selected. For this reason, an overview of different allocation procedures is given below.

An Overview of Different Allocation Procedures

The concept of apportioning global resources has been most widely researched and debated with respect to the emissions of carbon dioxide (CO_2). It is widely agreed that we should not exceed 1.5 °C of average global temperature increase (UNFCCC 2017). This limit can be translated into a total global "budget" of CO_2 emissions. For example, to have a 90% chance of limiting global warming to 1.5 °C total, global, cumulative emissions from 1870 onwards should remain below 555–615 GtC[1] (Rogelj et al. 2015). In 2011 we had already emitted 500 GtC (Stocker et al. 2013), leaving 55–115 GtC to be shared between today's population and future generations.

There are many different theories on who should have the rights to these remaining emissions. Some of the most commonly discussed allocation procedures include:

- Equal per capita share—each person on the planet has an equal right to the remaining allowable carbon emissions.
- Equal per capita share with historic accountability—each person who has lived or will live has an equal right to the total cumulative allowable carbon emissions.

---

[1] There is 1 kg of carbon in every 3.67 kg of carbon dioxide emitted. A budget of 500GtC equates to a 1835GtCO$_2$.

- Grandfathering—the rights to carbon emissions are based on a share of carbon emissions taken at a past reference date.
- Contraction and convergence—high users reduce emissions, while low users increase emissions until convergence at an equal per capita share level.
- Common but different—rights to the resource based on level of development already achieved.

Equal Per Capita Share: A Discussion

Equal per capita share, put simply, means that the global carbon budget is divided by the world population—everyone has the right to the same amount of emissions. This approach was proposed in an early draft of the International Framework Convention (INC 1991), a revised version of which was accepted at the 1992 Rio Earth Summit (Beckerman and Pasek 1995). Typically, this amount is then multiplied by each national population, and budgets are managed at this level, although there have also been proposals for a personal carbon budget (Lövbrand and Stripple 2011).

The main arguments for an equal per capita share are that it is fair:

- Every human being has an equal right to Earth's resources.
- That it is the only solution that will be widely accepted (Beckerman and Pasek 1995).
- That there is no duty to take on obligations resulting in the actions of one's ancestors (Caney 2013).
- That there is no international law or precedent that actions with unforeseen, unintended consequences must be mitigated (Beckerman and Pasek 1995).
- There is no objective way to calculate the costs and benefits of historic emissions and thus determine the appropriate compensation to or from each nation (Grubb 1995).

The counterarguments are that since current generations are benefitting and continue to benefit from past emissions, these beneficiaries should bear the burden of the environmental consequences (Shue 1999). Low-income countries argue that they should not be penalized for historical emissions by high-income countries (Ha and Teng 2013). They contend that low-emission countries will be incentivized to make financial gains from the system by selling their "excess" emission rights to high-emission countries who may be unable to make the cuts required to meet their budgets.

The arguments against this method are generally based on historical inequities. The carbon budget is cumulative, and past emissions make up the vast majority of the total. Countries which have more emissions in the past generally have more wealth, better infrastructure, and often a higher quality of life. Countries with low historic emissions are often facing the challenge of aspiring to the level of infrastructure and quality of life experienced in other countries but needing to do so without emitting similar quantities of carbon that were emitted by these wealthier countries.

Further, the global population are already experiencing negative side effects from past emissions. The planet has already warmed almost 1 °C (Stocker et al. 2013).

The countries experiencing the brunt of the climate impacts to date are often countries with very low past emissions. They are thus faced with the combined challenge of bearing the costs of climate adaptation and trying to develop without further emissions.

When considering a global carbon budget, the problem is less straightforward. There is a cumulative budget of emissions from 1870 onwards, and in 2017, much of this budget has already been used up (Stocker et al. 2013).

Those who advocate such an allocation argue the lifestyle experienced by those in high-income countries can be largely attributed to past emissions of carbon. Roads, power plants, buildings, and public transport systems take substantial amounts of energy to develop. In countries where this has already been developed, this was almost certainly at the cost of high levels of CO_2 emissions.

Emerging nations such as India and China have aspirations to bring living standards up to the same level as high-income nations but are faced with trying to do so in a time where we understand the importance of limiting carbon emissions on a global level.

Not every allocation method makes sense for every Quota. The methods above have been developed for carbon and GHGs for which a total budget over time exists. One of the greatest debates around the allocation of a global carbon budget is that historic emissions come from the same total budget as future emissions. This is not the case for every Quota. Most of the Quotas are based on annually renewing budgets—thereby altering the frame of the problem. In the example of nitrogen, we have exceeded the Planetary Boundary, and over the past decade, we have also exceeded the Planetary Quota each year. However, unlike the Carbon Quota, the Nitrogen Quota is still a positive number, i.e. we do not need to remove nitrogen from the system in order to return to the Planetary Boundary level—the nitrogen limit is not cumulative. As such, it does not make sense to apply the grandfathering approach in the way that one would for carbon.

This does not mean that the only approach that can work is the equal per capita approach. Rather, that new approaches should be considered and developed based on the different framing that Planetary Accounting provides. It is not necessary that one approach be applied across every Boundary. Some allocation approaches may consider regional differences for Quotas with high regional variability such as the Water Quota. By definition, the basis for the Planetary Accounting framework should have *different mechanisms at different scales*.

It should be considered that for such a flexible approach, the approach to allocation would also need a high degree of flexibility. A Quota for the basis of self-organized initiatives is likely to be self-selected. Global negotiations for national commitments to Quotas are likely to be heavily influenced by politics. Private organizations could agree a sectorial approach to setting Quotas, could self-select Quotas as part of an internal sustainability strategy, could have Quotas allocated by local authorities, or could report against Quota ranges in a similar way to the Netherlands study (Lucas and Wilting 2018).

Table 18.3 Divisible   and
nondivisible Quotas

Divisible quotas	Nondivisible quotas
Carbon	Air quality
MeNO	Biodiversity
Forest	
Ozone	
Nitrogen	
Phosphorous	
Water	

18.3.2.3 Calculating a Share of the Planetary Quotas

The Quotas as shown in Table 18.1 are global limits. Although each Quota is scalable, not every Quota should be scaled by division. The Carbon Quota is an example of a Quota that is divisible—i.e. the global Quota of -7.3 GtCO$_2$/yr. could be divided by the global population (say 7.5 billion) to get an equal per capita share of -1 tCO$_2$/yr. per person.

In contrast, the PQ for aerosols does not need to be divided to apply to different scales. The unit (aerosol optical depth equivalent) applies directly at any scale. Thus, the global Planetary Quota is the same as (for example) any individual's Planetary Quota. Table 18.3 shows which Quotas are divisible and which are not.

The share of the PQs determined can be viewed as an end goal. It is a policy, governance, or behavioural decision to determine how quickly the PQs should be met and how to do this. This does not define the pathways.

18.3.3 The Impact Balance Sheet

The results of the environmental impact assessment can then be compared to the scaled PQs in the Planetary Accounts. An "impact balance statement" can be used to show the impact and limit for each PQ currency and thus the credit or deficit.

18.3.4 Potential Applications

The accounts can then be used in any number of ways. They could inform policy and behaviour change; they could be used to compare impacts of different individuals, cities, products, or nations. They could be used as the basis for an international trading scheme.

18.3.4.1   Planning

Planetary Accounting could be used to assess the relative impacts of different future scenarios. Currently, it is possible to assess the relative impacts of different projects. Results of past environmental impact assessments might be able to tell us that a train line would reduce carbon impacts by 1000 t/yr and water impacts by 500 m^3/yr compared to new building codes which reduce carbon by 500 t/yr and water by 1000 m^3/yr. The decision-makers are then faced with a problem—which is more important. The Planetary Accounting Framework does not include a mechanism to rank different environmental currency. However, with Planetary Accounting, these numbers can be put into the context of scientific limits. A citywide impact balance statement might show that the city is doing pretty well against its PQ for water but has a long way to go to return to its PQ for carbon. As such, the city could make an informed decision to prioritize the train line. On the other hand, a different city might be struggling to meet its PQ for water, doing well on its PQ for carbon. In this city, the best environmental choice might thus be to amend the building code.

18.3.4.2   Policy

Scaled PQs can be used to determine future science-based targets and to understand the current impacts against these. This information would help to determine specific policies and pathways to reach the targets in a given timeframe. Science-based targets are becoming increasingly popular in the development of policy but are typically limited to carbon, water, and land. The PQs can be used to determine a full suite of science-based targets for critical Earth-system function.

18.3.4.3   Business Operations

The impact balance sheet and scaled PQs could also be used to inform business operations, sustainability strategies, and long-term planning. The scaled PQs are also useful in risk management. It is widely understood that businesses that emit carbon as part of their core business are at high risk of going out of business. The Quotas will allow businesses to speculate on future risks. For example, nitrogen and phosphorous use are not currently taxed. However, the PBs and PQs show that at a global scale, we are using dangerously high levels of these substances. This suggests that these substances could be taxed or limited in the future, information which might flag alarm bells early to allow businesses to begin working on reducing reliance on nitrogen and/or phosphorus before any limits or taxes are put in place.

18.3.4.4   Behaviour Change

Planetary Accounting could also be used to develop behaviour change applications, such as a smart phone app which allowed players to compete to reduce their own impacts against their PQs, and to compete against others (this is discussed further in Sect. 18.6).

18.3.4.5   Comparison Studies

Cities could use consumption-based accounting practices with an agreed scope to compare per capita impacts in different cities across the globe. The results may be very useful for collaborative efforts to reduce impacts. Investigations could be undertaken to understand why some cities have lower impacts than others. The results could help the cities to share ideas and cross-pollinate—one of the key benefits of a poly-scalar approach.

C40 Cities, a voluntary collective of mayors, is a great example of how this could work. The cities involved are working together to reduce carbon emissions at a city scale. Planetary Accounting would allow these efforts to be broadened across critical global variables.

18.3.4.6   Legislation

In the same way that Planetary Accounting could be used to inform policy, it could also be used as the basis for environmental legislation at different scales. Common Home of Humanity (CHH) is proposing a new framework for global environmental legislation based on the "condominium model".[2] In a condominium, different units are owned and essentially operated by different parties. However, there are legal frameworks in place to manage the shared systems such as the building envelope, roof, and plumbing and electrical systems. If there is an electrical problem, the collective owners have the right to access any part of the system to repair it, even if this access is within a privately owned space. Likewise, although internal walls where the electrical wires are housed may be privately owned, the owner does not have the legal right to alter or damage the wires as these are collectively owned.

CHH propose that a similar legal construct could be used to manage the Earth system. In the current global model, there are mechanisms to legislate in shared territory—the oceans and outer space. However, many elements that are critical to the overall functioning of the Earth system are located within national territories. The proposed condominium model would give collective rights to such elements. For example, the Amazon rainforest is essential to global wellbeing; however, it is currently managed by one nation.

---

[2] This would be called the "strata" model in Australia.

The CHH condominium model, with Planetary Accounting as the scientific basis for the global legislation, was shortlisted in the top 14 of thousands of entries to the Global Challenges Foundation New Shape Prize (Global Challenges Foundation 2017).

18.3.4.7   Product Information

To facilitate better producer and consumer responsibility, a product labelling system similar to the nutritional facts labelling system for food could be developed based on the PQs. Whether this was displayed on products as part of a labelling system, or simply made available online, companies could use such a system to communicate the impacts of goods and services in different environmental currencies. A global labelling scheme could provide an opportunity to address the regional variation of some PQs (such as the water Quota). This is discussed in more detail in Sect. 18.6.

18.4   Discussion

18.4.1   Timeframe

The Quotas represent the same safe operating state as the Planetary Boundaries—as such they refer to an end goal rather than a pathway of reductions. The purpose of the Quotas is to allow humanity the freedom and flexibility to determine the best way to operate within the safe operating space. There is no specific date before which the Quotas must be respected. At any time that any of the Quotas or Boundaries are not respected, humanity is at risk of an irreversible departure from a Holocene-like state.

18.4.2   Comparing Quotas

There is no mechanism to compare one PQ to another or to amalgamate the results of environmental assessments into a single indicator of sustainability. This is intentional. Earth cannot amalgamate environmental currencies such as carbon and water or trade one for another. If humans consume too much water, this cannot be resolved by emitting less carbon, though it is appreciated that there is a nexus between water and carbon. At a global scale, each of the PQs must be respected if we are to return to and operate within the Planetary Boundaries.

This does not preclude the opportunity to trade in each of the Quota currencies at lower scales. On the contrary, Planetary Accounting provides an opportunity for a global trading system for key global environmental "currencies", and in the process, firms can see how these parameters interact and are synergistic. Moreover, the real costs to humanity of exceeding planetary limits could be used to assign a monetary value to each environmental currency. For example, the costs of adaptation and mitigation of exceeding the PQ for nitrogen could be used to assign a monetary value per kg of nitrogen. Such an exercise could facilitate the incorporation of the environmental impacts into existing global economic frameworks, thus enabling a further developing of wealth creation and environmental footprint (Newman et al. 2017).

18.4.3   The Quotas Are a Moving Target Not a Static Value

The Earth system is dynamic and the rate of increase in scientific understanding of its processes and limits is high. There is not time to wait until we have a perfect understanding of the system or its limits before we take action to operate within these—this may never eventuate. The indicators and limits presented in this paper are intended to be preliminary. It is my intention that, like the Planetary Boundaries, these are subjected to scrutiny, discussion, and analysis and are regularly reviewed and updated over time as we advance in our collective knowledge and understanding.

18.4.4   Global vs Regional Limits and Impacts: An Issue of Scale

Carbon emissions are fundamentally different to most other planetary limits. Greenhouse gases have a long atmospheric lifetime and become well mixed in the atmosphere. This means that it is of little importance where the gas is emitted. 1 kg of CO_2 will have the same contribution to global warming wherever it is released.

When we consider other limits, for example, water consumption or the release of nitrogen into the environment, it is not the case that 1 kg consumed or released in one location will have the same impacts as 1 kg consumed or released elsewhere. If we take a few thousand litres of water from a water source with abundant supply, the local impacts are likely negligible. Taking just a few litres from another, water poor source, may have disastrous local effects. The release of a kilogram of nitrogen in a sparse agricultural area will have less impact on the Earth system than in an intense agricultural zone with risks of groundwater contamination.

One way to include regionality in Planetary Accounting could be through a product and services labelling scheme as identified previously. To give an example of how this could work, a binary water scarcity indicator (yes/no) could be reported alongside the net water footprint to convey the suitability of the water source. In the same vein, regional issues for other environmental currencies could be included in such a system—the release of aerosols has more impact in highly populated areas or areas that already suffer from air pollution than in areas where the air is clean. This information could also be included in a product labelling system.

In a similar manner, it would be interesting to explore the use of a binary efficiency indicator against a given benchmark. This would help put the raw environmental currency data into context for consumers. A tick or star system could be used to convey whether the results are better or worse than similar products.

Planetary Accounting is not intended as the one supersystem to resolve all environmental problems though it will contribute to most. The purpose of Planetary Accounting is to allow humanity to manage human activity such that it does not push the Earth system into a new geological state. There are many local environmental problems that do not significantly impact planetary limits. For example land instability and polluted waterways due to poor farming practices, light pollution, and urban heat island effects. Planetary Accounting does not replace local environmental management practices created locally and solvable locally; these must be dealt with at a local level.

This does not mean that regionality should be ignored. Regionality might be included in reporting planetary impacts through testing in demonstrations at different scalar levels appropriate to each of the Planetary Quotas.

18.5   Strengths and Weaknesses

The strength of this research lies in its combination of management theory, environmental accounting, and Earth-system science. Drawing on these fields allowed it to present a framework for evaluating environmental impact which was both scientifically rigorous and actionable. Previous research in this field (Rockström et al. 2009; Steffen et al. 2015; Sandin et al. 2015; Fang et al. 2015; Dao et al. 2015; Nykvist et al. 2013; Ewing et al. 2010) has been either scientifically rigorous or actionable, but not both.

A key element of this research was the identification of the importance of the type of indicator selected for policy applications. This insight has been identified by practitioners in the past. Yet it does not seem to have made its way to the field of Earth-system science. It is likely that the specifics of the Planetary Quotas may change once made available to the scientific community to assess and develop. However, the fundamental premise, that limits must be communicated in Pressure indicators, should not.

One of the strengths and weaknesses of the book is in its breadth. The task of connecting three previously unconnected fields of research and deriving Planetary Quotas across nine different fields of science would be a daunting task for a multidisciplinary team. To attempt this as a PhD student with a background in engineering and sustainable building design and a Professor of Sustainability with abilities in transport and city planning was overwhelming to say the least and could not have been possible without the help of many scientists who were able to explain their work and followed us down the track of delivering a Pressure indicator for the PBs.

The multidisciplinary aspect of the work can be considered a strength. The research connects science with policy and business. As shown in this book, this is a gap that needs to be addressed if we are to generate the magnitude of change needed. A weakness is that it was not undertaken by a multidisciplinary team who would likely have had additional insights that could not be determined without depth in each field. The extended peer community engagement was undertaken to address this weakness as far as possible, and indeed the scientific aspects of translating PBs into PQs were probably sufficiently rigorous due to the remarkable access to so many global leaders in this field. However, without time or budget, there were limits as to the involvement of other experts who could have been engaged in the process of creating actionable indicators. Perhaps other specialists in poly-scalar change management would have come up with a better approach, and time will tell if this now can happen.

The Planetary Accounting Framework (Fig. 18.3) outlines key steps that would apply to any level of application of the framework. Each of these steps will need to be developed to give a more detailed framework or perhaps several frameworks (e.g. to address the differences between applications for individuals, businesses, governments, and products).

Another limitation of Planetary Accounting is that it has not yet been applied and evaluated as an instrument to guide policy, business, or behavioural decisions. In the development of the concept, and particularly of the framework, much effort was taken to envision the different applications to determine and address potential weaknesses of the system. However, there is no substitute for real-world applications. Efforts are underway with the Planetary Accounting Network (see Sect. 18.6) to launch a pilot study.

The ten indicators selected for the Planetary Quotas vary in their robustness and in the likely availability of quality data. The indicator for the carbon PQ is already widely used which means that it is likely to be quite robust with plenty of data. In contrast, the indicator selected for the biodiversity PQ is relatively new, and the indicator for the aerosols PQ was developed as part of this thesis so data will not yet be ready yet. These indicators will need to undergo substantial testing in different applications to assess their robustness. It is likely that the data needed to measure impacts against these indicators will be difficult to find at first but not impossible.

18.6   Ongoing and Future Work

The Planetary Quotas and Planetary Accounting Framework were developed as a way to manage human impacts on the environment in order to return to the safe operating space defined by the Planetary Boundaries. There is much work to be done for this to become a reality.

There has already been some advancement of the work presented here. A study assessing the interaction between two of the PBs using the Planetary Accounting Framework and the Landau-Ginzburg model has recently been completed. The findings of this study were that the interaction studied was small, although nonvanishing. The authors argue that repeating this approach across the PBs would lead to a Quota-based accounting system that closely resembles Planetary Accounting.

Each of the Planetary Quotas should be reviewed and revised as needed by the scientific community. This should not be a one-off occurrence. Rather, this should happen on an ongoing basis, so that the limits reflect the latest scientific knowledge.

The Planetary Accounting Framework should be further developed. In particular, the System of Environmental and Economic Accounting Central Framework (SEEA-CF) should be used to inform this development to align the framework with international environmental reporting standards.

Further research should be undertaken to determine how best to communicate the PQs and PAF to the wider population. Feedback from the extended peer community engagement has already included the opinion that the nomenclature and scientific units used for the PQs are still too complex for the general public. This may be simply be improved by people involved in the next phase of application.

As described above, there is work underway to develop a legal framework for the Earth system, which would use Planetary Accounting as the scientific basis (see Sect. 18.3.4). This proposal is promising. However, more research should be undertaken to determine whether there are also other legal frameworks, governance structures, community engagement initiatives, or business drivers that could help to generate global change using a PQ approach.

One of the key benefits established for a poly-scalar approach to global environmental management is that it would enable trial and error of different solutions at different scales and under different circumstances. Findings from such varied approaches could help to accelerate change. However, this is only true where the lessons can be captured and shared. Research should be done to determine whether this could be possible through an online platform using means such as crowdsourcing. This has been done successfully to gather data in the past (see Box 18.1 for an example).

The research presented in this thesis has also highlighted areas of future work for the PBs. As identified in this thesis, the PBs are not in a uniform category of indicator. Some are Pressures, some are States, and one is an Impact (see Sect. 6.3). The role of the PBs is to assess and communicate planetary health. This is best done through State and Impact indicators. Where the PB indicators are Pressures, for example, water consumption, nitrogen fixation, or phosphorous release to the oceans, the health of the planet with respect to these processes is not well communicated. In the same way that the PBs using State and Impact indicators could be translated to pressure indicators, it would be possible to translate the PBs which are in Pressure indicators to States or Impacts. For example, the PB for water consumption could be translated to an indicator such as the percentage of water bodies experiencing water scarcity, and nitrogen and phosphorous indicators could be translated to global areas of aquatic dead zones.

Thus far, the PBs have been developed by a self-selected group of scientists. However, there is talk of developing an independent committee tasked with the ongoing management and updating of the PBs. If this was to occur, the same committee could also be tasked with linking these PBs to PQs in an ongoing manner. A full suite of State and Impact Boundaries, with a full suite of Pressure Quotas, that were maintained by an independent scientific body, would be a powerful tool for Earth-system management. It is not hard to imagine how such a group could generate a global research and policy process similar to that developed by the IPCC for climate change. This process is led by several thousand scientists sharing their findings with policy-makers in an ongoing dialogue. The processes that began through the IPCC have created significant change (Newman et al. 2017) though much still needs to be done. Such a process could now be shifted to include the Planetary Boundaries and show the synergies and trade-offs that could be created by bringing all the PQs together into a PAF that is constantly being updated and demonstrated.

18.6.1 The Planetary Accounting Network

On the back of this research, Kate Meyer launched a new not-for-profit research centre called the Planetary Accounting Network (PAN), with Peter Newman as the founding member and advisor (PAN 2019). PAN is a New Zealand-based, global, member-based organization which undertakes evidence-based research in the new field of Planetary Accounting to help people and organizations to operate within the planet's limits. PAN was launched in October 2018 and has already had exceptional support from a variety of New Zealand and international organizations and global experts. It was launched to continue and advance the work presented in this book.

There are several projects currently being targeted at PAN. These include:

- A detailed Planetary Accounting Protocol that provides guidance for those seeking to undertake Planetary Accounting at an individual, business, city, or national scale.
- A pilot study to test the protocol at these scales.
- The derivation of economic value of key environmental impacts, using the PQ "currencies".
- The gamification of the PAF in a smart phone application.
- A "planetary facts" labelling system for products and services.

These ideas are expanded below.

18.6.1.1 The Planetary Accounting Protocol

PAN is seeking funding to develop a detailed accounting protocol that will define how to undertake Planetary Accounts at four key scales. The protocol will include scope and boundary definitions, data collection procedures, calculation methodologies, and assumptions and simplifications needed where data gaps exist. This work will be undertaken with extensive peer community engagement—in particular with stakeholders and end users related to the four scales.

18.6.1.2 The Planetary Accounting Pilot

PAN is currently in discussion with a number of organizations regarding launching a pilot study at key scales as proof of concept. Pilot study participants will be involved in the development of the protocol but will engage PAN as an independent environmental accounting practitioner to ensure transparency and robustness through this study.

18.6.1.3 Economic Value

Scientists can estimate the cost of mitigating and adapting to environmental degradation. For example, there have been estimates of the social cost of CO_2 which range from USD\$12–64 per tonne (IWGSCC 2013). This cost is based on future damages avoided derived from predicted costs from impacts such as sea level rise and changes in agricultural yields and ecosystem function. There is a high degree of uncertainty in such estimations. However, assigning costs to environmental impacts begins to communicate the importance of these in a global language. Further work should include the continued development of social cost estimates for environmental currencies, and the estimation of the true value of each of the PQ currencies, per unit of impact.

Eventually these costs will be seen as real and the need for governments to intervene and create regulatory frameworks that would seek to reduce these costs. One approach is to create a market approach for each of the PQs. Major polluters would need to pay the true costs associated with their impacts, and these would be carried over to consumers of such impacts. Countries with environmental assets critical to Earth-system functioning could be financially incentivized to maintain these. It could form the basis of initiatives such as a globally capped impact trading scheme which could happen at any scale of activity across all of the PQ currencies. The inclusion of all PQ environmental impacts into the existing economic structures would constitute systemic change and show how the growth in each PQ can be decoupled from growth in wealth as is happening with greenhouse emissions (Newman 2017).

18.6.1.4 Gamification of the PAF

There are several personal impact calculators available online (e.g. Global Footprint Network 2018; WWF 2018; Anthesis 2014; n-print 2012; Water Footprint Network 2018). These allow users to calculate their impacts such as their ecological, water, nitrogen, or carbon footprints. There has been rapid growth in the number of online personal impact (PI) calculators over recent years suggesting increased interest in personal sustainability (Franz and Papyrakis 2011).

The purpose of these calculators is to educate players and encourage behaviour change to reduce impacts (Franz and Papyrakis 2011). However, a review of popular online calculators shows that most calculators:

- Propose limits that are not based on scientific planetary limits.
- Do not propose limits at all.
- Where limits are proposed, do not provide options which allow players to win, i.e. even when the best options are selected, the impacts shown exceed the proposed limits.
- Are based on average per capita national production impacts, i.e. the data is skewed to show higher impacts for people living in countries which are net exporters than for people in net importing countries.
- Use generic impacts/$ to estimate impacts of goods and services.

The result can be that players are left with a sense of confusion and/or doom (see Fig. 18.4). Most online PI calculators do not encourage behaviour change. They may therefore be quite counterproductive.

Planetary Accounting could be used as the basis to advance personal impact calculators and develop "real-life" games that mean behaviour change and potential productive actions at many scales. It could be used as the basis for the impacts assessed and end goal targets proposed. Engagement with game developers would be needed to determine the best way to design the game to generate a high uptake of users. Further research would be required to determine the most effective ways to generate change through games. The idea of using games to change behaviour has proven successful in the past. For example, SPARX (S mart, P ositive, Active, R ealistic, X -factor thoughts) is a game which has been shown to reduce teenage depression scores as successfully as cognitive behaviour therapy delivered by a qualified psychologist (Merry et al. 2012).

18.6.1.5 Planetary Facts Labelling

To facilitate better producer and consumer responsibility, a product labelling system similar to the nutritional facts labelling system for food could be developed based on the PQ (see Fig. 18.5). Whether this was displayed on products as depicted or made available in some other way would need to be determined. The communication of the impacts in each PQ currency and the proportion of a recommended PQ that the product comprises would enable consumers to begin to understand the impacts of their purchasing decisions in the context of global limits. This is fundamentally different from existing labelling schemes which typically provide information about impacts compared to industry benchmarks.

18.7 Concluding Remarks

Generating change to live within planetary limits is more difficult than simply knowing what these limits are. It is necessary to understand how people behave and what drives people to make certain choices and what structures create dependence

Fig. 18.4 A cartoon depiction of the failure of many online personal impact calculators to achieve their fundamental goal – to improve individual behaviour

Nutrition Facts

Serving Size 2/3 cup (55g)
Servings Per Container About 8

Amount Per Serving

Calories 230   Calories from Fat 40

	% Daily Value*
Total Fat 8g	**12%**
Saturated Fat 1g	**5%**
Trans Fat 0g	
Cholesterol 0mg	**0%**
Sodium 160mg	**7%**
Total Carbohydrate 37g	**12%**
Dietary Fiber 4g	**16%**
Sugars 1g	
Protein 3g	

Vitamin A	10%
Vitamin C	8%
Calcium	20%
Iron	45%

* Percent Daily Values are based on a 2,000 calorie diet.
  Your daily value may be higher or lower depending on
  your calorie needs.

	Calories:	2,000	2,500
Total Fat	Less than	65g	80g
Sat Fat	Less than	20g	25g
Cholesterol	Less than	300mg	300mg
Sodium	Less than	2,400mg	2,400mg
Total Carbohydrate		300g	375g
Dietary Fiber		25g	30g

Planetary Facts

Serving Size 2/3 cup (55g)
Servings Per Container About 8

Amount Per Serving

	% Daily Value*
Carbon 8g CO_2	**60%**
Ag-GHGs 4g CO_{2e}	**5%**
CH_4 2.5g CO_{2e}	**3%**
N_2O 1.5g CO_{2e}	**2%**
Nitrogen 2g N_r	**20%**
Phosphorous 2kg P	**3%**
Deforestation -0.2 m^2	**14%**
Aerosols 0.01 AUD_e	**40%**
Water 20kg H_2O	**16%**
Biodiversity 0.18E-14 PDF	**100%**
Ozone 0 ODPkg MGs	**100%**
Imperishable Waste 3kg	**3000%**

*Percent Daily Values are based on an daily average of the
equal per capita share for a 7.5 billion population. Annual per
capita share listed below

	Impact	
Carbon	Less than	-0.3tCO_2
Ag-GHGs	Less than	0.72tCO_{2e}
Nitrogen	Less than	8.2kg
Phosphorus	Less than	1.5kg
Deforestation	Less than	-1.4m^2
Aerosols	Less than	0.1AOD_e
Water	Less than	5.3m^3
Biodiversity	Less than	1.8E-14
Ozone	~	0 ODP kg/yr
Imperishable Waste	~	0 kgs

Fig. 18.5 Planetary Facts labels could give consumers information in the same way that nutrition facts tell consumers what is in their food

on the actions which lead to impacts. Further, one must consider current environmental management practices and the advantages and limitations of these. To generate serious change so that we can live within the planet's environmental limits requires integrative thinking that brings together the scientific knowledge of Earth's limits, the utility of environmental impact assessment frameworks, and the understanding of behaviour, change, and management theories. This book has begun to show how this can be done. The research presented here shows how the Planetary Boundaries can be translated into Planetary Quotas and the Planetary Accounting Framework to make global environmental limits accessible and actionable to all scales of human activity. This approach could form the basis for the management of the Earth system to help us to return to and live within the Planetary Boundaries, the "safe operating space" for humanity.

References

Anthesis (2014) what is your ecological footprint [Online]. Available: http://ecologicalfootprint.com/. Accessed 28 April 2018

Beckerman W, Pasek J (1995) The equitable international allocation of tradable carbon emission permits. Glob Environ Chang 5:405–413

Caney S (2013) Justice and the distribution of greenhouse gas emissions. *Global Social Justice.* J Glob Ethics 5(2):125–146

Dao H, Peduzzi P, Chatenoux B, de Bono A, Schwarzer S, Friot D (2015) Environmental limits and Swiss footprints based on planetary boundaries. UNEP/Grid-Geneva & University of Genever, Geneva, Switzerland

Dawkins E, Moran D, Palm V, Wood R, Bjork I (2019) The Swedish footprint: a multi-model comparison. J Clean Prod 209:1578–1592

Ewing B, Moore D, Goldfinger S, Ourslet A, Reed A, Wackernagel M (2010) Ecological footprint atlas 2010. Global Footprint Network, Oakland

Fang K, Heijungs R, de Snoo GR (2015) Understanding the complementary linkages between environmental footprints and planetary boundaries in a footprint-boundary environmental sustainability assessment framework. Ecol Econ 114:218–226

FAO (2016) Global forest resources assessment. NATIONS, F. A. A. O. O. T. U, Rome, p 2015

Franz J, Papyrakis E (2011) Online calculators of ecological footprint: do they promote or dissuade sustainable behaviour? Sustain Dev 19:391–401

Global Challenges Foundation (2017) New shape prize [Online]. Available: https://globalchallenges.org/our-work/the-new-shape-prize. Accessed 3 March 2019

Global Footprint Network (2018) Footprint calculator [Online]. Global footprint network. Available: http://www.footprintnetwork.org/resources/footprint-calculator/. Accessed 28 April 2018

Grubb M (1995) Seeking fair weather: ethics and the international debate on climate change. Int Aff 71:463–496

Ha Y, Teng F (2013) Midway toward the 2 degree target: Adequacy and fairness of the Cancún pledges. Appl Energy 112:856–865

Hoekstra AY (2017) Water footprint assessment: evolvement of a new research field. Water Resour Manag:1–21. https://doi.org/10.1007/s11269-017-1618-5

INC (1991) Framework convention on climate change. Intergovernmental Negotiating Committee

IWGSCC (2013) Technical update of the social cost of carbon for regulatory impact analysis US government

Kaza, Silpa, Yao L, Bhada-Tata P, Frank VW (2018) What a waste 2.0 : a global snapshot of solid waste management to 2050. World Bank, Washington, DC

Lövbrand E, Stripple J (2011) Making climate change governable: accounting for carbon as sinks, credits and personal budgets. Crit Policy Stud 5:187–200

Lucas P & Wilting H (2018) Towards a safe operating space for the Netherlands: Using planetary boundaries to support national implementation of environment-related SDGs

Merry SN, Stasiak K, Shepherd M, Frampton C, Fleming T, Lucassen MFG (2012) The effectiveness of SPARX, a computerised self help intervention for adolescents seeking help for depression: randomised controlled non-inferiority trial. Br Med J 344:e2598

N-PRINT (2012) Your nitrogen footprint [Online]. Available: http://www.n-print.org/YourNFootprint. Accessed 28 April 2018

Newman P (2017) Decoupling economic growth from fossil fuels. Mod Econ (8):791–805

Newman P, Beatley T, Boyer H (2017) Resilient cities: overcoming fossil fuel dependence. Island Press/Center for Resource Economics

Nykvist B, Persson Å, Moberg F, Persson LM, Cornell SE, Rockström J (2013) National environmental performance on planetary boundaries: a study for the Swedish environmental protection agency. AGENCY, S. E. P, Sweden

PAN (2019) The planetary accounting network [Online]. Available: www.planetaryaccounting. org. Accessed

Rockström J, Steffen W, Noone K, Persson A, Chapin FS III, Lambin E, Lenton TM, Scheffer M, Folke C, Schellnhuber HJ, Nykvist B, de Wit CA, Hughes T, van der Leeuw S, Rodhe H, Sörlin S, Snyder PK, Costanza R, Svedin U, Falkenmark M, Karlberg L, Corell RW, Fabry VJ, Hansen J, Walker B, Liverman D, Richardson K, Crutzen P, Foley J (2009) Planetary boundaries: Exploring the safe operating space for humanity. Ecol Soc 14:32

Rogelj J, Luderer G, Pietzcker RC, Kriegler E, Schaeffer M, Krey V, Riahi K (2015) Energy system transformations for limiting end-of-century warming to below 1.5 °C. Nat Clim Chang 5:519–527

Sandin G, Peters GM, Svanström M (2015) Using the planetary boundaries framework for setting impact-reduction targets in LCA contexts. Int J Life Cycle Assess 20:1684–1700

Shue H (1999) Global environment and international inequality. Int Aff 75:531–545

Steffen W, Richardson K, Rockström J, Cornell SE, Fetzer I, Bennett EM, Biggs R, Carpenter SR, de Vries W, de Wit CA, Folke C, Gerten D, Heinke J, Mace GM, Persson LM, Ramanathan V, Reyers B, Sörlin S (2015) Planetary boundaries: guiding human development on a changing planet. Science 347:1259855

Stocker TF, Qin D, Plattner G-K, Alexander LV, Allen SK, Bindoff NL, Bréon F-M, Church JA, Cubasch U, Emori S, Forster P, Friedlingstein P, Gillett N, Gregory JM, Hartmann DL, Jansen E, Kirtman B, Knutti R, Krishna Kumar K, Lemke P, Marotzke J, Masson-Delmotte V, Meehl GA, Mokhov II, Piao S, Ramaswamy V, Randall D, Rhein M, Rojas M, Sabine C, Shindell D, Talley LD, Vaughan DG, Xie S-P (2013) Technical summary. In: Stocker TF, Qin D, Plattner G-K, Tignor M, Allen SK, Boschung J, Nauels A, Xia Y, Bex V, Midgley PM (eds) The physical science basis. Contribution of working group I to the Fifth assessment report of the intergovernmental panel on climate change. Cambridge University Press, Cambridge, United Kingdom and New York, NY, USA

UN (2016) Montreal protocol–submission to the high-level political forum on sustainable development (HLPF) 2016. United Nations Sustainable Development Knowledge Platform: UN

UNFCCC (2017) The Paris agreement [Online]. Available: http://unfccc.int/paris_agreement/ items/9485.php. Accessed 19 Sep 2017

Water Footprint Network (2018) Personal water footprint [Online]. Available: http://waterfootprint.org/en/water-footprint/personal-water-footprint/. Accessed 28 April 2018

WHO (2016) Ambient air pollution: a global assessment of exposure and burden of disease. In: World Health Organisation. WHO Press, Geneva, Switzerland

World Bank (2019) CO_2 Emissions (metric tons per capita) [Online]. Available: http://data.worldbank.org/indicator/EN.ATM.CO2E.PC?order=wbapi_data_value_2009%20wbapi_data_value%20wbapi_data_value-last&sort=asc. Accessed 18 May 2019

WWF (2018) How big is your environmental footprint? [Online]. WWF. Available: http://footprint.wwf.org.uk/. Accessed 28 April 2018